Green Energy and Technology

Climate change, environmental impact and the limited natural resources urge scientific research and novel technical solutions. The monograph series Green Energy and Technology serves as a publishing platform for scientific and technological approaches to "green"—i.e. environmentally friendly and sustainable—technologies. While a focus lies on energy and power supply, it also covers "green" solutions in industrial engineering and engineering design. Green Energy and Technology addresses researchers, advanced students, technical consultants as well as decision makers in industries and politics. Hence, the level of presentation spans from instructional to highly technical.

Indexed in Scopus.

Indexed in Ei Compendex.

Philip Pong
Editor

Renewable Energy Resources and Conservation

Springer

Editor
Philip Pong
Electrical and Computer Engineering
New Jersey Institute of Technology
Newark, NJ, USA

ISSN 1865-3529 ISSN 1865-3537 (electronic)
Green Energy and Technology
ISBN 978-3-031-59004-7 ISBN 978-3-031-59005-4 (eBook)
https://doi.org/10.1007/978-3-031-59005-4

This Springer imprint is published by the registered company Springer Nature Switzerland AG
The registered company address is: Gewerbestrasse 11, 6330 Cham, Switzerland

Contents

Part II Clean Energy Technology and Emission Reduction

Part III Waste-to-Energy and Microbial Fuel Cell Technology

Part I
Solar Energy and Photovoltaic Power Generation

Artificial Neural Network Application for the Prediction of Global Solar Radiation Inside a Greenhouse

Salah Bezari, Asma Adda, Sofiane Kherrour, and Reda Zarrit

1 Introduction

Solar energy is an amazing and sustainable resource that has the potential to revolutionize the way to meet various sectors' energy needs, especially in the greenhouse agricultural. The climate has a direct impact on agricultural productivity, particularly in greenhouse farming where a favorable microclimate is essential for good production. In this context, solar energy can be harnessed to power these structures and create a sustainable solution for the agricultural sector. For example, there is research on improving greenhouse systems by storing large-scale solar thermal energy to heating [1]. Renewable energy solutions like solar power are truly revolutionizing the way we meet our energy needs. It is, indeed, the irregular distributions of temperature, relative humidity, carbon dioxide concentration and solar radiation in the microclimate of greenhouse that can cause negative impacts on growth and quality. Nowadays, artificial neural networks are used in many applications of renewable energies such as desalination [2, 3], conditioning of greenhouses [4, 5], and solar energy [6, 7]. Several studies have been carried out on greenhouses, for example, to control the climate of the greenhouse [8], the thermal modeling of the climate of the greenhouse [9], and the prediction and estimation of the climatic factors of the greenhouse [10]. The local greenhouse climate can be predicted through knowledge of physics and information derived from data on inside and outside variables. It seems that Ferreira et al. [11] conducted a study on the use of RBF neural networks to model the indoor air temperature

S. Bezari (✉) · S. Kherrour · R. Zarrit
Unité de Recherche Appliquée en Energies Renouvelables, URAER, Centre de Développement des Energies Renouvelables, CDER, Ghardaïa, Algeria

A. Adda
Laboratory of Biomaterials and Transport Phenomena (LBMPT), Faculty of Science and Technology, University of Dr Yahia Fares, Medea, Algeria

P. Pong (ed.), *Renewable Energy Resources and Conservation*, Green Energy and Technology, https://doi.org/10.1007/978-3-031-59005-4_1

of a greenhouse. S. Zeng et al. [12] utilized several input parameters release to predict the temperature and humidity inside greenhouses. The results showed that the proposed model exhibited better adaptability and more satisfactory real time. R. Ahmad and his team [13] used a neural network model with the Levenberg-Marquardt algorithm to predict temperature and humidity in a greenhouse with natural ventilation. Kuzugudenli, E. [14] conducted a study in 2018 where they developed regression and artificial intelligence network models to predict relative humidity in greenhouses using monthly mean temperature, total precipitation, and altitude parameters from 177 weather stations in Turkey. In 2020, Escamilla-García et al. [15] conducted a review of artificial neural network (ANN) applications in greenhouse technology. This study developed a model based on an ANN for predicting incident solar radiation on a horizontal surface in an agricultural greenhouse.

2 Materials and Methods

2.1 Experimental Greenhouse

The experimental greenhouse, which is the subject of this work, is placed within the Applied Research Unit for Renewable Energies (URAER) located in the south of Algeria, city of Ghardaïa with latitude: +32.37°, longitude: +3.77°, and altitude of 450 m above mean sea level (see Fig. 1). This greenhouse is equipped with a set of sensors to measure the temperature, relative humidity, and solar radiation.

2.2 Modeling Procedure

In our study, neural networks were chosen to address greenhouse modeling, as previous studies have shown them to be useful and powerful tools to define such

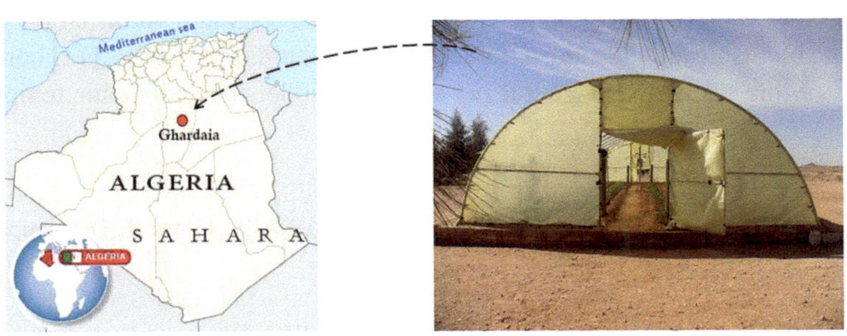

Fig. 1 Greenhouse view in URAER site (Ghardaïa region)

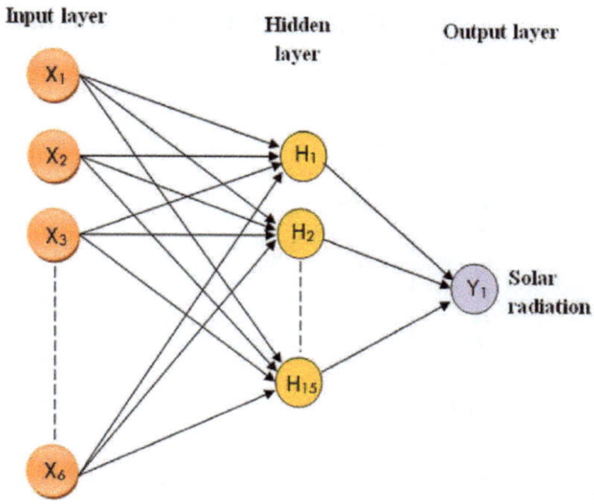

Fig. 2 ANN architecture

a system [4]. Solar radiation plays a major role in the climate of the greenhouse; it therefore constitutes the network output of the model studied. In this way, the model will try to predict the solar radiation incident on a horizontal surface inside the greenhouse depending on the variables used as input to the network, such as the temperature of the outside and inside air, the relative humidity, solar radiation, etc. Regarding the data sets used, the sampling time to obtain the measurements was 30 min, which corresponds to 145 measurement points in 72 h. The database used contains a monthly set of parameter values for 3 days for the month of the winter season (January). A Multilayer Perceptron (MLP) consists of at least three layers named input layer, output layer, and hidden layers (see Fig. 2).

To get the most accurate estimate of our model's performance, it is important to calculate the error. The best way to do this is by using statistical parameters such as the determination coefficient R, the root mean square error RMSE, and mean absolute error MAE. These parameters are frequently referenced in literature and can be calculated using specific equations outlined in references [2].

3 Results and Discussion

For the studied network, the Levenberg-Marquardt (LM) learning algorithm with 15 neurons in the hidden layer for the network (7-15-1) produced the best results, and it is used to generate the graphical output. The mean squared error (MSE) during the training of three phases (training, validation, and testing) is shown in Fig. 3. The backpropagation learning error plot explains that the error is high when the iteration is less and vice poured.

To some extent, errors in training, validation, and test sets indicate how well the performance of the trained network can be measured. Therefore, it is necessary to

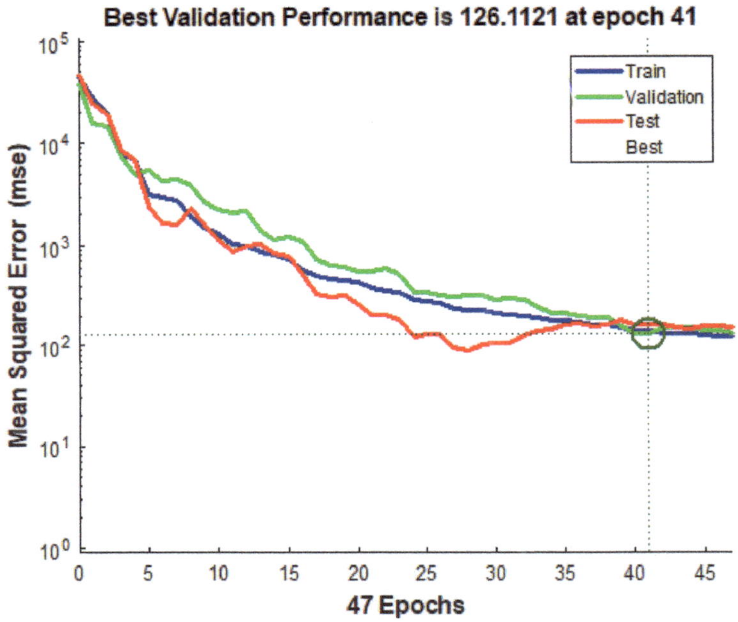

Fig. 3 Diagram of MSE as function the number of iteration (epochs)

study the response of the network by performing regression analysis, which is a measure of how the variance of the results is explained by the objectives.

The R values obtained for the target output from each estimated division are: (training 99.95%, validation 99.58%, test 99.32%, and overall 99.78%, as shown in Fig. 4. Through the regression, the ANN made good accuracy.

Figure 5 presents the comparison between the measured and predicted solar radiation inside the greenhouse using the MLP neural network. A comparison made between the results of the measured values and those observed according to the ANN model for the entire database shows a good correlation, which results in a coefficient of determination of 0.99. The model developed has a dynamic close to the system in the learning phase. On the other hand, in the test phase, the error seems significant even if overall the dynamics are respected and the estimated output presents some "peaks." The latter is due to the influence of the input elements on the solar radiation to the front of the greenhouse.

Table 1 presents the validation agreement plot for the inside greenhouse solar radiation with an agreement vector approaching the ideal $[\alpha, \beta, R] = [1.008, -0.0791, 0.997]$.

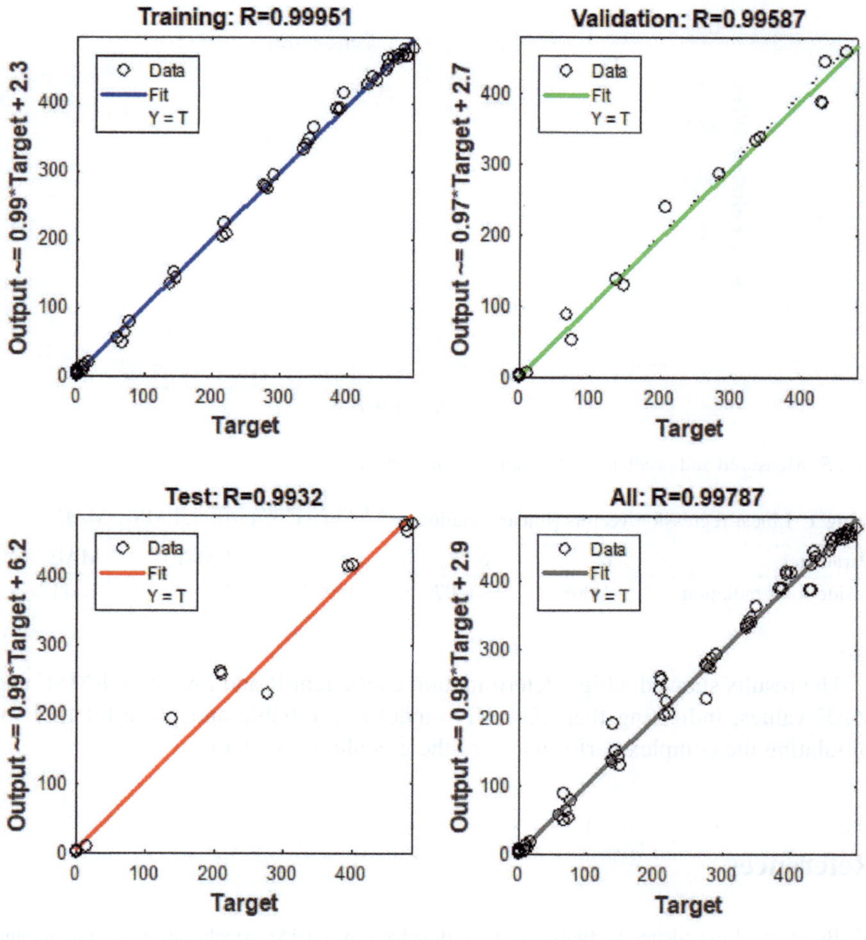

Fig. 4 Regression result of ANN training. (**a**) training, (**b**) validation, (**c**) test, and (**d**) all

4 Conclusions

Over the past decades, research and studies in greenhouse engineering have focused on reducing production costs and the need to reduce environmental impacts to ensure production quality.

In this study, we relied on a tunnel-type agricultural greenhouse covered with plastic film in a desert area in Ghardaïa. The greenhouse is equipped with several sensors and a climate station to record various parameters.

To manage and control the performance of solar greenhouse, an ANN model was developed to predict the solar radiation on a horizontal surface inside the greenhouse.

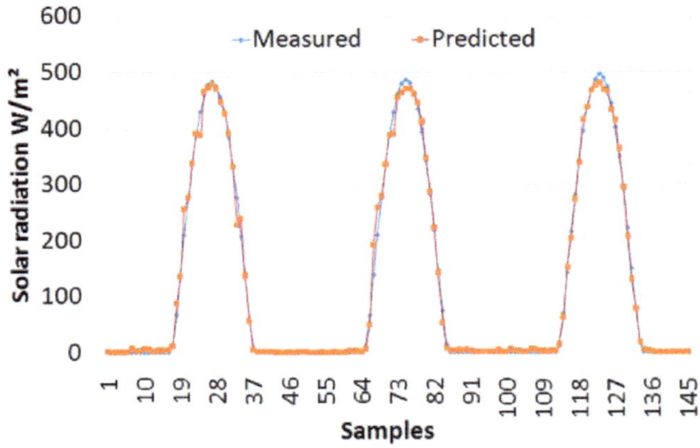

Fig. 5 Measured and predicted solar radiation in greenhouse

Table 1 Linear regression vectors [linear equation: $y^{\text{predict}} = \alpha y^{\text{exp}} + \beta$], R, RMSE, MAE

Parameter	α	β	R	RMSE	MAE
Inside solar radiation	1.008	−0.079	0.997	1.56	1.311

The results showed a high determination coefficient R and low error RMSE and MAE values, indicating that the ANN model is a reliable and powerful tool for simulating the complex performance of the greenhouse system.

References

1. Bezari, S., Bekkouche, A., Bensaha, H., & Benchatti, A. (2015). Amelioration of a greenhouse through energy storage system case study: Ghardaia region. In *International conference on renewable energy research and applications* (pp. 578–582). IEEE.
2. Adda, A., Hanini, S., Bezari, S., Ameur, H., & Maouedj, R. (2020). Managing and control of nanofiltration/reverse osmosis desalination system: Application of artificial neural network. *International Journal of Design & Nature and Ecodynamics, 15*(6), 843–853.
3. He, Q., Zheng, H., Ma, X., Wang, L., Kong, H., & Zhu, Z. (2022). Artificial intelligence application in a renewable energy-driven desalination system: A critical review. *Energy and AI, 7*, 100123.
4. Taki, M., Ajabshirchi, Y., Ranjbar, S. F., & Matloobi, M. (2016). Application of neural networks and multiple regression models in greenhouse climate estimation. *Agricultural Engineering International: CIGR Journal, 18*(3), 29–43.
5. Belouz, K., Nourani, A., Zereg, S., & Bencheikh, A. (2022). Prediction of greenhouse tomato yield using artificial neural networks combined with sensitivity analysis. *Scientia Horticulturae, 293*, 110666.
6. Siham, C. M., Salah, H., Maamar, L., & Latifa, K. (2017). Artificial neural networks based prediction of hourly horizontal solar radiation data: Case study. *International Journal of Applied Decision Sciences, 10*(2), 156–174.

7. Barrera, J. M., Reina, A., Maté, A., & Trujillo, J. C. (2020). Solar energy prediction model based on artificial neural networks and open data. *Sustainability, 12*(17), 6915.
8. Belalem, M. S., Elmir, M., Tamali, M., Mehdaoui, R., Missoum, A., Chergui, T., & Bezari, S. (2021). Numerical and experimental study of natural convection in a tunnel greenhouse located in South West Algeria (Adrar region). *International Journal of Heat and Technology, 39*(5), 1575–1582.
9. Aissa, M., & Bezari, S. (2018). The orientation effect of the agricultural tunnel greenhouse on aerodynamic and energy properties. In *5th international symposium on environment-friendly energies and applications* (pp. 1–4). IEEE.
10. Mohammed, R., & Allal, S. (2022). The prediction of the inside temperature and relative humidity of a greenhouse using ANN method with limited environmental and meteorological data. In *E3S Web of conferences* (Vol. 351, p. 01004). EDP Sciences.
11. Ferreira, P. M., Faria, E. A., & Ruano, A. E. (2002). Neural network models in greenhouse air temperature prediction. *Neurocomputing, 43*(1–4), 51–75.
12. Zeng, S., Hu, H., Xu, L., & Li, G. (2012). Nonlinear adaptive PID control for greenhouse environment based on RBF network. *Sensors, 12*(5), 5328–5348.
13. Ahmad, R. O. B. I. A. H., Lazin, M. N. M., & Samsuri, S. F. M. (2014). Neural network modeling and identification of naturally ventilated tropical greenhouse climates. *WSEAS Transactions on Systems and Control, 9*(1), 445–453.
14. Kuzugudenli, E. (2018). Relative humidity modeling with artificial neural networks. *Applied Ecology & Environmental Research, 16*(4), 5227–5235.
15. Escamilla-García, A., Soto-Zarazúa, G. M., Toledano-Ayala, M., Rivas-Araiza, E., & Gastélum-Barrios, A. (2020). Applications of artificial neural networks in greenhouse technology and overview for smart agriculture development. *Applied Sciences, 10*(11), 3835.

Prototype of a Solar Photovoltaic Charging Station Applied to the Propulsion of Artisanal Fishing Vessels in Arequipa, Peru

Juan José Milón Guzmán, Mario Enrique Díaz Coa, Jorge Antonio Molina Díaz, and Diego Alonso Valdivia Vera

1 Introduction

Currently there is a trend on concepts of electromobility as a measure that has been proposed to decarbonize transportation systems worldwide. This new concept of the use of electric motors brings with it various benefits such as the positive impact on the environment due to the elimination of polluting emissions, greater energy efficiency of all systems that use electric motors and lower acquisition costs. It is for these reasons that in various parts of the world the use of electric motors as part of electromobility is being promoted through policies, guidelines, and standards [1]. This new trend has been applied to various sectors such as marine transport. In Norway, for example, the propulsion of zero-emission marine vessels has been promoted by electrifying them and this has reduced consumption costs of fossil fuels with close to 98% of the electricity used coming from renewable energy. However, various limitations have been verified, such as the lack of charging stations to facilitate the autonomy of the vessels, which is why charging stations have been implemented in the ports [2]. Artisanal fishing in Peru is a sector that faces great problems related to the successful development of its activities. Nearly 90% of the artisanal vessels use gasoline as fuel in two-stroke engines, which is why one of the main expenses incurred is fuel, amounting to as much as 40 USD for a day's work. It has been identified that one of the needs in terms of extraction capacity and cost reduction is the use of electric propulsion systems [3]. It has since been verified that this type of electric motor with a power of 10 kW can mobilize a vessel supplying a combustion engine of up to 60 HP [4]. However, the use of such electric motors has a major weakness, which is the lack of recharging infrastructure for battery banks, since not all places, such as remote locations where access to electricity by network

J. J. Milón Guzmán (✉) · M. E. Díaz Coa · J. A. Molina Díaz · D. A. Valdivia Vera
Universidad Tecnológica del Perú, Lima, Peru
e-mail: jmilon@utp.edu.pe

P. Pong (ed.), *Renewable Energy Resources and Conservation*, Green Energy and Technology, https://doi.org/10.1007/978-3-031-59005-4_2

is limited, have a charging station. H. Wang et al. [5] investigated the potential benefits of applying solar panel systems to a Ferry, for which they calculated the life cycle of photovoltaic systems as a viable, economical, and environmentally sound solution to replace traditional systems such as diesel engine systems. It was determined thereby that the recovery period of the investment in the solar panel system is only 3 years. Y. D. Çağlar [6] evaluated the contribution of solar energy in marine vessels by designing solar panels on a Cargo Ship with a total length of 208.3 m, a hybrid system consisting of two diesel generators, 1274 solar panels, and an inverter. The photovoltaic system covered 38% of the fuel requirement of the vessel, which is equivalent to 73.53 T of fuel with an energy efficiency of 7.76%, and a recovery period of 7.2 years. In Ecuador, D. Umarani et al. [7] proposed to install a solar charging station in the open spaces of an educational campus in order to provide a constant power supply to mobile devices and laptops in open spaces using clean energy from the sun. Meanwhile, M. Shubham et al. [8] studied the charging of electric bicycles in workplaces through an off-grid system with a battery bank. In that study, a simulation was carried out using the Matlab software, and later the scenarios with a voltage of 12 V were compared to 24 V and 36 V, respectively. Peña and Céspedes [9] proposed the design of a photovoltaic solar charging station in a shopping center and undertook a technical-economic analysis of the charging station for electric vehicles. Over the years the demand for energy has increased, especially since the consumption of electricity is now used for mobility in ocean-going vessels [10, 11].

2 Experimental Model

The experimental model (Fig. 1) is made up of two battery banks, a storage system, a solar supply system, a charge control system, an electrical supply system, and a measurement and data acquisition system. The battery bank is made up of two movable metal structures that contain four batteries each, providing for a total of eight batteries. The structure of the battery bank has fast connectors that facilitate the connections between both banks, charge and discharge connections that will be used in the experimental tests. The batteries that were used are $LiFePO_4$, each of the batteries has an approximate weight of 32 kg, Model LIT 100-48S, Ultracell, 48 V, 100 A·h, 4800 W·h. These batteries were connected in parallel. The charge controller is the model MPPT 250|100, Victron Energy brand, 5800 W. The solar panels used are 370 W, 48 V, and 22 kg. A 5000 VA Victron Energy Multiplus 48|5000|70 model charger was used to charge the battery banks with the electricity grid and with an electric generator. This charger also serves as an inverter. Two Panther K0E4 4 kW electric generators (Inverter generator) and the 6 kW Yamaha EF7200DE generator (Inverterless generator) were used, both fueled by gasoline. Hioki CM7290 brand clamps were used for current measurement. The Data Acquisition System is a Keysight brand model DAQ970A.

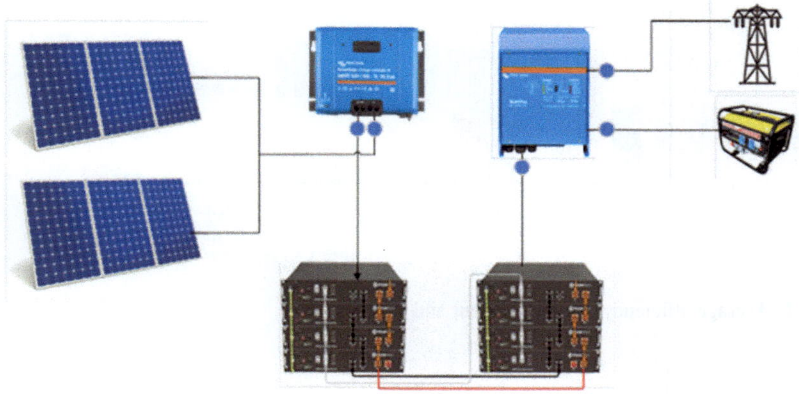

Fig. 1 Setup of the experimental model

The studied uncertainties [12] are Current $\pm$ 2%, Voltage $\pm$ 0.15%, Power $\pm$ 2%, Fuel mass $\pm$ 0.25%, and Irradiance $\pm$ 0.25%.

The Efficiency of the solar photovoltaic system is:

$$\eta_{\text{solar}} = \cdot I/I_S \cdot A \tag{1}$$

V = Voltage [V]; I = current [A]; Is = Solar irradiance [W/m^2]; A = Area [m^2]

3 Results

The tests were conducted in September and October 2022 (Spring), with a peak solar irradiance of 1150 W/m^2 (Arequipa, Peru). The following graph details the charging time for each of the charging types for one bank (four batteries) and two battery banks (eight batteries). It should be noted that in the case of the charging time of the solar panels, the total time has been counted, that is, including the night hours when there is no presence of the sun. It can be highlighted that the types of load that have a shorter charging time are generators and the electrical network (Fig. 2).

Figure 3 shows the efficiencies calculated for four batteries with different charging methods. A great variation can be observed between each of the methods, the most efficient being the charge with the eight photovoltaic panels, which is close to 99% followed by the efficiency with four photovoltaic panels, with a lower efficiency. The load with the electrical network can also be observed, which is more stable and constant since a linear trend is displayed, finally the load efficiency with the generator can be visualized.

Figure 4 shows the efficiencies calculated for eight batteries with different charging methods. A great variation can be observed between each of the methods,

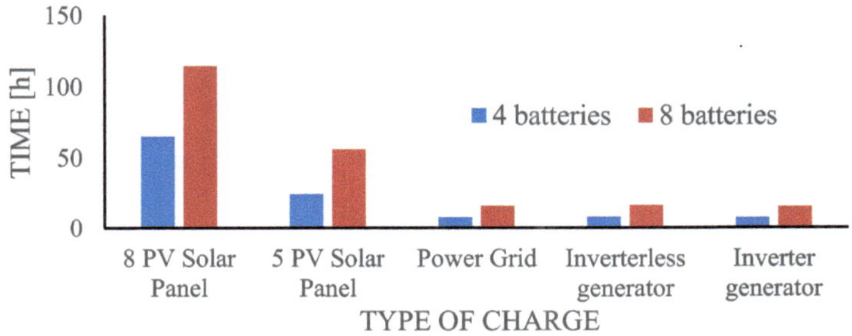

Fig. 2 Average efficiency of charging four and eight batteries

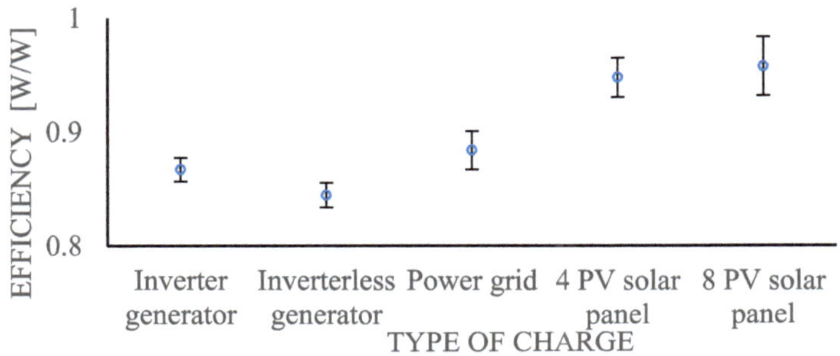

Fig. 3 Battery charge time according to type of charge

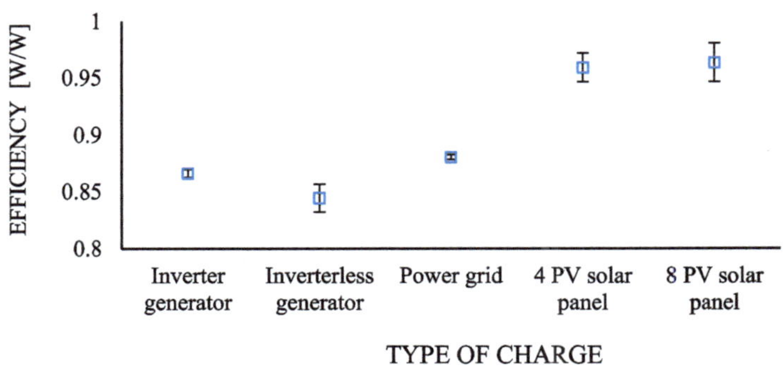

Fig. 4 Average efficiency of charging eight batteries

the most efficient being the charge with the eight photovoltaic panels, which is 96.4% followed by the efficiency with four photovoltaic panels that is 95.9%. It can also be seen that the uncertainties of the generators are less than those obtained by the four and eight photovoltaic panels. It can also be affirmed that the uncertainty of the electrical network is the lowest. It is for this reason that the efficiencies with uncertainties are very close to the average. For charging a battery bank, the Inverter generator was more efficient than the inverterless generator because the average power generated is closer to the rated power of the equipment, unlike the inverterless

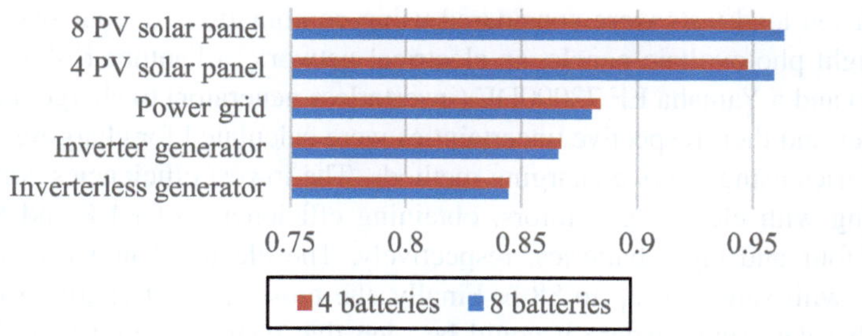

Fig. 5 Comparison of average efficiencies for four and eight batteries

generator whose efficiency decreased because the average power was further away from that of the rated power of the equipment.

Figure 5 presents a summary of the average charging efficiencies for four and eight batteries using the five previously mentioned charging methods. It can be highlighted that the lowest efficiencies correspond to the charge with generators, the charge with the inverterless generator being the lowest, which is close to 84% for the charge of four and eight batteries, compared to the charge with the Inverter generator, which is much more efficient since it represents an 86% average efficiency. It can also be seen that the electrical network is even more efficient and stable, with an average efficiency close to 88%. However, the most efficient means of charging is that of the solar panels where it can be seen that charging eight batteries with eight panels is the most efficient with 96.3% compared to charging with four panels where a slightly lower efficiency is obtained at 95.8 %.

Two (2) discharge tests of a battery bank (made up of four batteries) and two battery banks (totaling eight batteries) were carried out. In these tests, the discharge system with two inverters and discharge module was used. The discharge time of the bank of four batteries was 6 h. The average discharge voltage was 48 V, the average discharge current was −64 A, and the average discharge power was −3 kW. The discharge time of the bank of eight batteries was 8.6 h. The average discharge voltage was 48 V and the average discharge current was −90.8 A. The average discharge power for eight batteries was −4.4 kW. For the discharge test with the electric boat motor, a support structure was built in the laboratory to allow the motor to be anchored. A pool was also used in which the motor was submerged. After the pool was filled, the battery and throttle were connected to the motor.

4 Conclusions

The charging station was dimensioned using a charger inverter, photovoltaic panels, a charge controller, connection cables, and two electric generators. A prototype electric station used to charge two battery banks was built. The prototype was equipped with current and voltage sensors connected to the data acquisition system.

A total of ten load tests were considered using combinations of four photovoltaic panels, eight photovoltaic panels, an electrical network, a Panther K0E4 (Inverter generator) and a Yamaha EF 72000DE (inverterless generator) to charge them. The efficiencies and their respective uncertainties were calculated for charging four and eight batteries using the five charging methods. The lowest efficiencies correspond to charging with electric generators, obtaining efficiencies of 84% and 86% for charging four and eight batteries, respectively. The electrical network improves efficiency with values of up to 88%. Finally, the most efficient charging methods were with solar panels where it could be seen that charging eight batteries with eight panels is the most efficient with 96.35% compared to charging with four panels where a slightly lower efficiency is obtained, 95.7%. The charging time with photovoltaic panels is much longer than using the grid or electric generators, but, in terms of charging costs and efficiency, they produced very favorable results. The charging costs of a bank and two banks of batteries were evaluated in the charging station prototype, where it can be concluded that the cost of charging with the generators is much higher with costs between 85 PEN and 210 PEN. On the other hand, the cost of charging the battery banks using the domestic network is cheaper at 32 PEN and 34 PEN with domestic electrical networks of Arequipa and Matarani, respectively.

References

1. Salazar Lopez, J. J., Torres, E. M. G., & Galarza, D. F. C. (2020). Recharge of electric vehicles through a mixed whole optimization with participation of the demand response. *Revista I+D Tecnológico, 16*(2), 94–100.
2. Karimi, S., Zadeh, M., & Suul, J. A. (2020). Shore charging for plug-in battery-powered ships: Power system architecture, infrastructure, and control. *Electrification Magazine, 8*(3), 47–61.
3. Intelfin Estudios & Consultoría. (2019). *Diagnóstico de la demanda de financiamineto del sector de pesca artesanal y de menor escala en el Perú*. WWF.
4. Pancha Ramos, J. M., Hidalgo, V. J. R., & Reinoso, E. V. R. (2020). Implementation of an electric motor for light river transport units. *Polo del conocimiento, 5*(6), 187–204.
5. Wang, H., Oguz, E., Jeong, B., & Zhou, P. (2019). Life cycle and economic assessment of a solar panel array applied to a short route ferry. *Cleaner Production*, 471–484.
6. Çağlar Karatuğ, Y. D. (2020). Design of a solar photovoltaic system for a Ro-Ro ship and estimation of performance analysis: A case study. *Solar Energy, 207*, 1259–1268.
7. Umarani, D., Seyezhai, R., Pavithraa, S., Priya, S. N., & Meenapriya, K. (2021). Design and implementation of solar docking station for smartphones/laptops. *Materials Today*, 1–6.
8. Shubham, M., Gaurav, D., Subho, U., & Anurag, C. (2021). Modelling of standalone solar photovoltaic based electric bike charging. *Materials Today: Proceedings*.
9. Peña Ramos, C., & Céspedes Gonzales, M. (2021). Design of a solar charging station for electric vehicles in shopping malls. *ENERLAC, V*(2), 134–155.
10. Kolodziejski, M., & Pozoga, I. M. (2023). Battery energy storage systems in ships' hybrid/electric propulsion systems. *Energies, 16*(3), 1–24.
11. Abu Bakar, N. N., Guerrero, J. M., Vasquez, J. C., Bazmohammadi, N., Yu, Y., Abusorrah, A., & Al Turki, Y. A. (2021). A review of the conceptualization and operational management of seaport microgrids on the shore and seaside. *Energies, 14*(23).
12. Moffat, R. (1988). Describing the uncertainties in experimental results. *Experimental Thermal and Fluid Sicence, 1*(1), 3–17.

Evaluation and Improvement of the Efficiency of a Self-Contained Photovoltaic System Applied to a Small Business in Arequipa, Peru

Juan José Milón Guzmán, Diego Andree Reynoso Yana, Holger Campos Paredes, and Jhonatan Orlando Macedo Luna

1 Introduction

For some years now, the issue of renewable energy and the necessary research for its efficient use has been widely discussed. The sun is an inexhaustible source of clean energy that is available to anyone who has the initiative to use it, but its use throughout the world is still low compared to conventional sources of power generation, such as fossil fuels. This scenario has caused PV costs to drop rapidly, averaging 13% and 18% per year between 2009 and 2014 for residential and nonresidential buildings, respectively. The unit cost of photovoltaic power has been reduced more than 20 times since 1973, going from 100 to 0.30 USD/Wp. These readjustments have allowed the use of this technology to become widespread and are now on the threshold of massive applications [1, 2]. A correct dimensioning of photovoltaic components determines that the highest efficiency of this can be achieved, although technical criteria such as the coupling of the consumption and solar power curves must be taken into account [3]. In commercial or industrial applications, it is possible to oversize the photovoltaic system without significantly affecting profitability, since a larger system can cover a possible increase in energy consumption in the future [4]. If the surpluses are very high, these will not contribute to the recovery of the investment, since there is no discharge to the public network [5]. According to Willborn [6], it is stated that within the commercial sector, small and medium-sized companies, as compared to larger commercial businesses, are the ones that can obtain a greater benefit in profitability from photovoltaic generation systems. This situation occurs because most of these companies have

J. J. Milón Guzmán (✉) · D. A. Reynoso Yana · H. Campos Paredes · J. O. Macedo Luna
Universidad Tecnológica del Perú, Lima, Peru
e-mail: jmilon@utp.edu.pe

P. Pong (ed.), *Renewable Energy Resources and Conservation*, Green Energy and Technology, https://doi.org/10.1007/978-3-031-59005-4_3

the bulk of their energy needs during the day. According to Herce [7], in a study applied in Portugal, the use of renewable resources for energy generation and its application under the category of self-consumption is very widespread in Europe, with the classic user of the energy system becoming a producer and consumer at the same time. Users under this category are now known as "prosumers." As explained by Ramírez [8], the application of photovoltaic generation systems is more feasible in the economic field if they are applied to commercial sectors. Grid parity, which is the equality of grid energy prices and the energy generated by the photovoltaic system would be easily achieved by up to 88% of the total of the commercial sector, making the economic indicators favorable. According to Vásquez [9], to carry out the dimensioning of a photovoltaic system, the historical consumption of electrical energy must be analyzed, based on which it is possible to determine if the system deserves the implementation of an energy storage system (batteries). In his own work, Peralta [10] states that the constant supervision of the operation of the photovoltaic system must be carried out together with the installation of sensors that record the operating parameters such as voltage and current to determine the efficiency of the system. The main restriction of the photovoltaic systems that operate under the self-consumption model is that of not being able to inject surplus power into the electricity grid, which means that there is a considerable percentage of electrical power that is not consumed by the installation and therefore is considered lost. This percentage in countries with considerable photovoltaic development can reach 20% of the total energy generated [11]. In order to obtain maximum efficiency from photovoltaic systems, prosumers can increase the level of photovoltaic self-consumption in their facilities. One strategy to achieve this increase in self-consumption is to store excess generated energy that is not immediately consumed by the utility installation. This storage can occur through batteries or in domestic water heating systems [12, 13]. Kumar [14] explains that another means of reducing power fluctuations is by load shedding or reduction of charge, but that energy storage remains as one of the most effective methods of handling intermittency in PV systems. According to Abou [15], energy storage systems have been successfully used in cooperation with photovoltaic systems in isolated microgrids. Energy storage is viable and power can be delivered back to the facility when needed. Bhayo [16] explains that the storage of energy in batteries and its integration with photovoltaic systems ensures that the installation has an energy backup system. This means that when the photovoltaic generation is not enough to feed a load and there is a deficit of power, this is provided from the battery system to meet internal demand. However, battery systems are still expensive and represent about 54% of the total cost of capital. Saini [17] also explains that battery systems store the excess power generated by the photovoltaic system, functioning as an alternative electrical load during a low demand situation. On the other hand, when they are discharged and deliver the stored energy, they behave as an additional generator during peaks in demand.

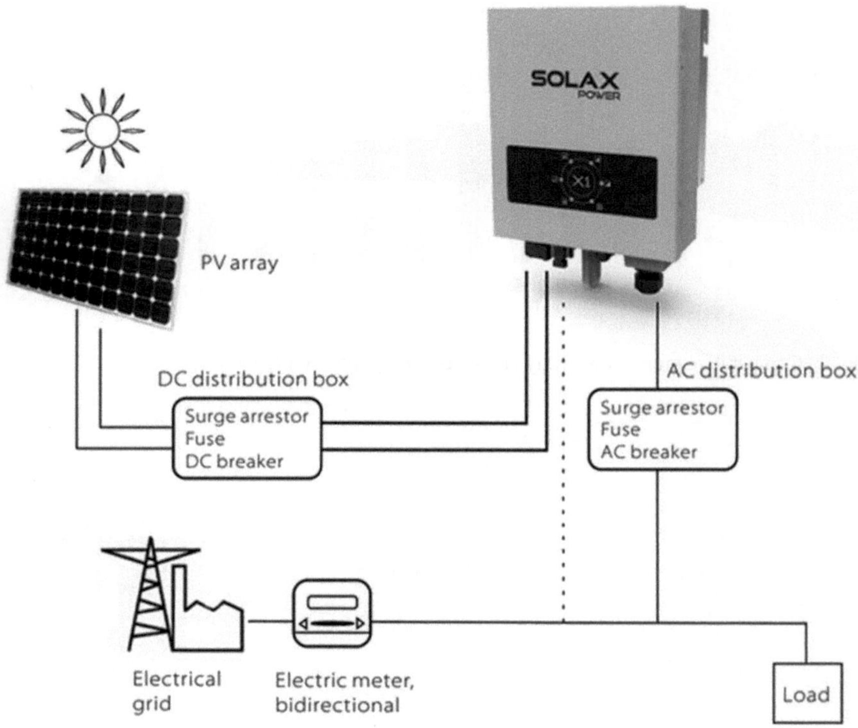

Fig. 1 Model of the photovoltaic system connection

2 Experimental Approach

The installation of the components that make up the photovoltaic solar system was carried out with the objective of obtaining the necessary data to determine the viability of said system. Figure 1 shown below shows the electrical connection model of the photovoltaic solar system and the existing electrical installation. The main components of the photovoltaic generation system can be identified, as well as the connection of the installation to the public electricity network.

3 Results and Analysis

Figure 2 shows the energy consumed by the installation, both the energy generated by the photovoltaic system, and that immediately consumed by the installation as it does not have storage systems; as there was energy consumed from the public network to satisfy the energy needs not covered by the photovoltaic system, comparative data can be obtained that will show the total values of energy effectively

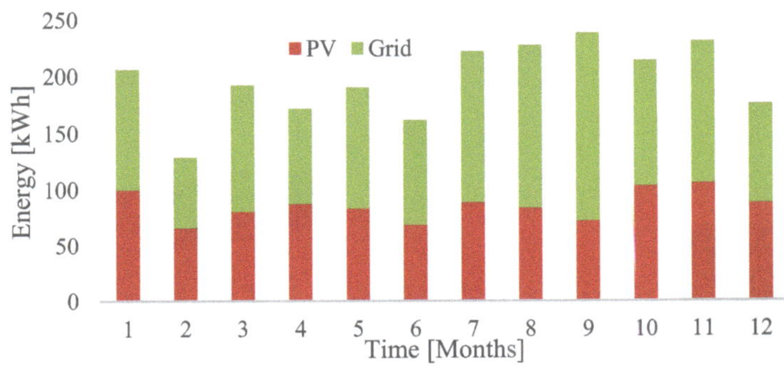

Fig. 2 Comparative graph of monthly energy consumption. (Source: Own)

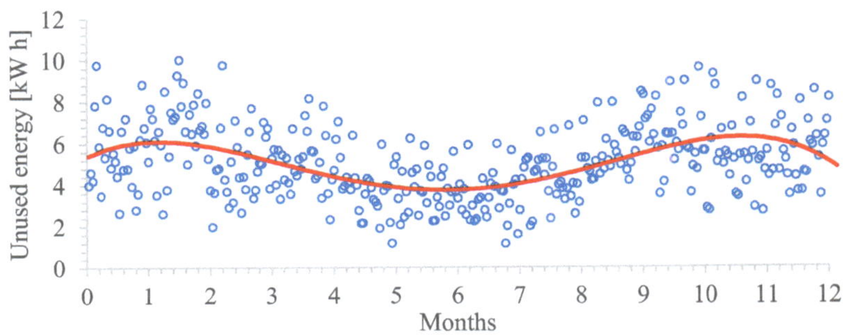

Fig. 3 Dispersion of annual data on unused energy

consumed by the installation. The annual value of the energy used for self-consumption was 1027.75 kW·h, while the annual value of energy that was taken from the public network was 1328.37 kW·h, making a total of 2356.12 kW·h

3.1 Process of Improvement

Figure 3 shows the values of wasted solar energy in the year. This trend cannot be linear due to the inconsistent patterns of solar radiation. For this reason, it is necessary to apply a polynomial trend. The result is seen as a curve in red, which also coincides with the behavior of solar radiation throughout the year. During the first 3 months, the energy is maintained at a high level, and from month four to month six the lowest energy value was obtained. This value then went on to recover its high value in the following months.

Figure 4 shows the data obtained from the quarterly averages. The lines that accompany the average data indicate the margin of error related to each of the averages.

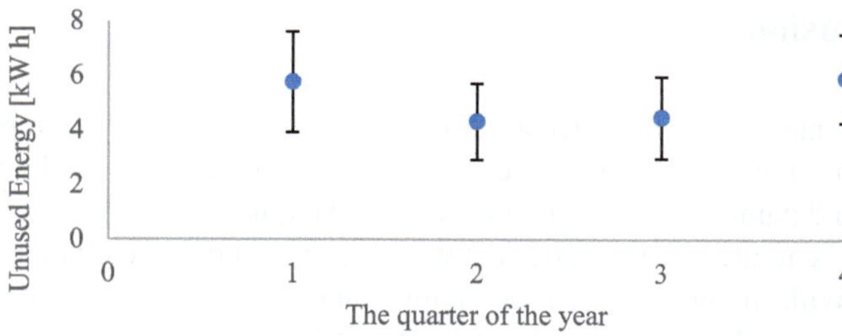

Fig. 4 Quarterly averages of unused energy

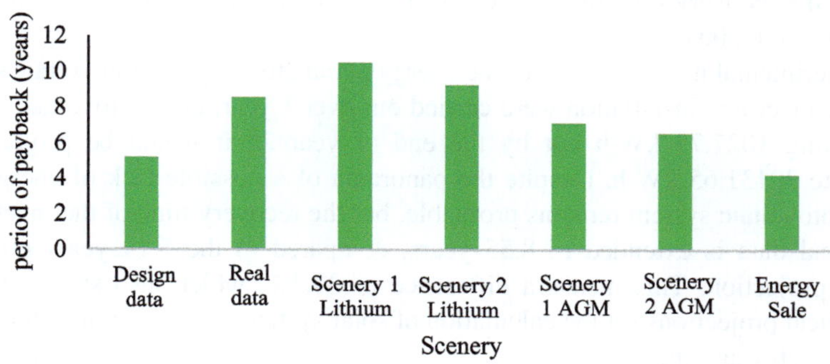

Fig. 5 Investment recovery period

Two different scenarios can be proposed for improvement (Fig. 5), a first scenario of maximum use of energy (scenario 1) and a second scenario that will prioritize the economy (scenario 2). In this scenario of maximum energy storage, it will seek to transform the greatest possible amount of solar radiation into electrical energy, and for this it is necessary to have a considerable amount of energy storage capacity so it can then be used to cover the needs of the installation when the photovoltaic system does not have the capacity to do so. To achieve the maximum use of solar energy, the average annual energy value obtained in Fig. 5 will be taken and, as the greatest amount of energy to be stored is sought, the margin of error is added, resulting in an average value of 6.89 kW·h of energy to be stored on a daily basis. In the priority scenario for economic savings, economic savings must be prioritized. For this purpose, the use of the minimum average amount of energy must be projected. In this case, the lowest average value of Fig. 5 must be taken and it is prioritized by subtracting the saving from the margin of error. In this way a value of 2.91 kW·h of energy to be stored with a battery system is obtained. Scenarios 1 and 2 propose the use of an energy storage system with different storage capacities. Based on these requirements, the use of two currently available technologies is proposed. The first is the use of lithium batteries and the second is to use AGM-type sealed lead-acid batteries as an alternative.

4 Conclusions

An experimental study was carried out to establish the efficiency of a photovoltaic solar system that is operating under the self-consumption model in a vehicle bodywork and painting workshop in the city of Arequipa, Peru. After evaluating the energy data generated by the system, it was determined that the efficiency achieved was 6.7%. With the projection of economic data, the system can be deemed to be profitable. Even when the projections based on linear averages of electrical loads and solar radiation are insufficient to satisfy the demand of a real work environment, it is necessary then to apply additional techniques that can improve energy use and reduce the recovery time of an investment so that these types of projects are more attractive to an investor.

Experimental measurements of the energy parameters of the photovoltaic system and the electrical installation were carried out over 1 year, the photovoltaic system generating 1027.75 kW·h but by the end of year 25 it would be projected to generate 30131.65 kW·h. Despite the panorama of a possible lack of energy use, the photovoltaic system remains profitable, but the recovery time of the investment with real data is extended to 8.55 years, compared to the 5.19 years proposed in the projection. This means a difference of 39.3%, which also shows that the theoretical projections for the calculation of solar systems are inaccurate and cannot be irrefutably assumed.

Different energy scenarios were presented where the use of two different battery bank technologies were proposed, using Lithium and AGM type batteries in order to store the energy wasted in times of low consumption of the installation while considering a maximum storage criterion of energy and a more conservative criterion from the economic point of view. With the first criterion, the use of solar potential can be improved to go from generating 1027.75 kW·h annually to generating 2091.21 kW·h, which would mean an increase of 203.5% in solar generation. With the second criterion, it goes from generating 1027.75 kW·h annually to generating 1877.34 kW·h, an increase of 182.66% in solar energy generation.

References

1. Elshurafa, A. M., Alsubaie, A. M., Alabduljabbar, A. A., & Al-Hsaien, S. A. (2019). Solar PV on mosque rooftops: Results from a pilot study in Saudi Arabia. *Journal of Building Engineering*, (25), 11.
2. Jiménez Castillo, G., Muñoz Rodriguez, F. J., Rus-Casas, C., & Talavera, D. L. (2019). A new approach based on economic profitability to sizing the photovoltaic generator in self-consumption systems without storage. *Renewable Energy*.
3. Milón Guzmán, J. J., Leal Braga, S., Zúñiga Torres, J. C., & Del Carpio Beltrán, H. J. (2020). Sizing methodology for photovoltaic systems considering coupling of solar energy potential and the electric load: Dynamic simulation and financial assessment. *E3S Web of Conferences*, (181).

4. Simola, A., Kosonen, A., Ahonen, T., Ahola, J., Korhonen, M., & Hannula, T. (2018). Optimal dimensioning of a solar PV plant with measured electrical load curves in Finland. *Solar Energy*, (170), 113–123.
5. Bastida Molina, P., Saiz Jiménez, J. Á., Molina Palomares, M. P., & Álvarez Valenzuela, B. (2017). Instalaciones solares fotovoltaicas de autoconsumo para pequeñas instalaciones. Aplicación a una nave industrial. *3C Tecnología: glosas de innovación aplicadas a la pyme, 6*(1), 1–14.
6. Willborn, S., Hesse, A., Balser, A., & Luh, A. (2014). *Study on the profitability of commercial self-consumption solar installations in Germany*. REC Solar Germany GmbH.
7. Herce Villar, C., Neves, D., & Silva, C. A. (2017). Solar PV self-consumption: An analysis of influencing indicators in the Portuguese context. *Energy Strategy Reviews*, (18), 224–234.
8. Ramirez Sagner, G., Mata Torres, C., Pino, A., & Escobar, R. (2017). Economic feasibility of residential and commercial PV technology: The Chilean case. *Renewable Energy, 108*, 20.
9. Vásquez Ducep, G. F., Artist. (2019). *Microgeneración distribuida con sistema fotovoltaico para autoconsumo en la Municipalidad de Picsi en el departamento de Lambayeque*. [Art]. Universidad Nacional Pedro Ruiz Gallo.
10. Peralta Vera, A. A., Artist. (2018). *Evaluación técnico-económica de una instalación de bombeo solar fotovoltaico aplicada a la ampliación de la frontera agrícola en zonas aisladas de Arequipa*. [Art]. Universidad Tecnológica del Perú.
11. Yuegu, W., Songsheng, Z., Jing, C., Zhaolin, W., & Song, H. (2018). Ammonia (NH3) storage for massive PV electricity. *Energy Procedia, 150*, 99–105.
12. Yildiz, B., Bilbao, J. I., Roberts, M., Heslop, S., Dore, J., Bruce, A., MacGill, I., Egan, R. J., & Sproul, A. B. (2021). Analysis of electricity consumption and thermal storage of domestic electric water heating systems to utilize excess PV generation. *Energy, 235*.
13. Say, K., Schill, W.-P., & John, M. (2020). Degrees of displacement: The impact of household PV battery prosumage on utility generation and storage. *Applied Energy, 276*.
14. Kumar, D. S., Maharjan, S., & Srinivasan, A. D. (2022). Ramp-rate limiting strategies to alleviate the impact of PV power ramping on voltage fluctuations using energy storage systems. *Solar Energy, 234*, 377–386.
15. Abou El-Ela, A. A., El-Seheimy, R. A., Shaheen, A. M., Wahbi, W. A., & Mouwafi, M. T. (2021). PV and battery energy storage integration in distribution networks using. *Journal of Energy Storage, 42*.
16. Bhayo, B. A., Al-Kayiem, H. H., & Gilani, S. I. (2019). Assessment of standalone solar PV-Battery system for electricity generation and utilization of excess power for water pumping. *Solar Energy, 194*, 766–776.
17. Saini, P., & Gidwani, L. (2022). An investigation for battery energy storage system installation with renewable energy resources in distribution system by considering residential, commercial and industrial load models. *Journal of Energy Storage, 45*.

Feasibility Study of a Sustainable Roof Top Domestic Solar Energy System in the UK

Ewen Constant and Joel Richards

1 Introduction

Work carried out prior to the global outbreak of Covid/SARS and the energy crisis that hit Europe and the world at large is revisited to understand if the changes in societal behaviour, such as increased home working being the new normal and the associated changes in energy consumption for the domestic user have had any effect on the payback periods for domestic home energy system via a solar photovoltaic system.

There have been many changes in behaviour for white collar workers in the UK due to Covid, home working is now not only expected by employees, but forms part of contracts written by organisations keen to reduce their footprint, and hence reduction in costs associated with maintaining empty offices. In the UK, and in Wales in particular, where this research is based, there has been a significant drop in the number of staff now attending offices on a full-time basis.

The Welsh government published figures in December 2022, which show only 10% of its staff are in the office at any point in time. These missing workers who are now home working have changed the pattern of energy consumption profiles across Wales. According to the Department for Business Energy & Industrial Strategy [1], UK domestic energy consumption has risen by 4% during weekdays, and for the first time in more than 12 years adjusted annual average domestic energy demand in the UK rose by approximately 200 kWh to 3954 kWh [2].

This is exacerbated by the drive by the UK government to halt the sale of all light and heavy fuel oil cars in the UK by 2035. The electric car market in the UK now accounts for 35% of all new vehicle sales, and places further demands on

E. Constant (✉) · J. Richards
University of South Wales, Pontypridd, Wales, UK
e-mail: ewen.constant@southwales.ac.uk

P. Pong (ed.), *Renewable Energy Resources and Conservation*, Green Energy and Technology, https://doi.org/10.1007/978-3-031-59005-4_4

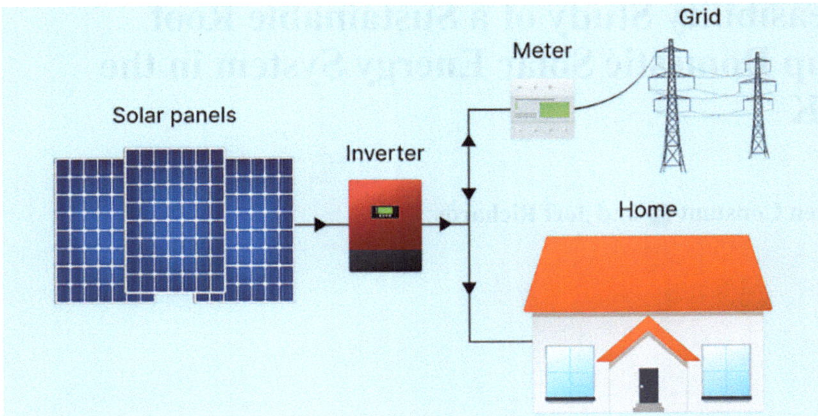

Fig. 1 Grid connectivity schematic [13]

the electricity network and is influencing consumer demand, which further effects demand profiles.

The previous work carried out [3–5] showed that there was an improved payback period by using a small-scale BESS facility linked to a PV system as shown in Fig. 1, as opposed to a standalone grid connected PV system. The three papers looked into not only domestic energy consumption, but also added the link to housing in remote locations, where off-grid living is not a desirable option for many users, but the main premise was to create a capture storage and reuse facility that reduces the need for grid fed electricity. This is a highly desirable option, removing the need to burn fossil fuels, and thus helping the global warming crisis by reducing the need for grid fed electricity. The work showed that cost-effective grid connected domestic home energy systems were available to consumers, but that due to the high cost of energy storage, payback periods were in the region of 20 years.

The review of this initial work aimed to prove a cost-effective system with reduced payback periods when considering grid connected domestic PV systems. It is clear since 2018 when energy prices in the UK were around 15 p/kWh in comparison to today's figure of around 35 p/kWh that energy costs have risen by almost two and a half times and that neither the energy storage solution, Lithium-Ion batteries, or the cost of installing a 4 kW system have risen by the same factor. To add to this amidst increasing energy prices it is reported by the institution for Fiscal Studies [6] that the average income fell by 1.7% during 2020–2021 in the UK.

2 Grid Tied Solar Systems

The most common type of photovoltaic technology used are monocrystalline and polycrystalline solar panels as well as thin-film solar cells. Each have advantages over one another in terms of efficiency rating at standard test conditions (STC), which is the rating given to a panel under the following setting:

- Irradiance measured at 1000 W/m^2.
- Ambient temperature of 25 °C.
- Air mass of 1.5.

Under these ideal standard test conditions, the panels will be given a maximum power rating in Watts. The efficiency of the panel can then be calculated using the formula:

$$\text{Efficiency\%} = \frac{P\max}{\left(\text{Area} \times 1000\ \text{W/m}^2\right)} \times 100 \tag{1}$$

The following data is based on a 4 KW solar installation using 16 * 250 W Solar Panels with an area of 1.4 m^2, a common size for 250 W panels. The use of PVGIS [7] has been used to collect average daily irradiation data per month of co-ordinates 51.405, −3.265 located in Barry, South Wales. This has assumed a fixed plane with an Azimuth of 0° degrees (α) facing south, with a slope incline of 35° (β). The raw global irradiation (G_i) data has a unit of w/m^2, therefore the total area of the panel system (S_{pv} * N) can be multiplied by the G_i data to give our values. The data was then adjusted to account for the efficiency losses (ζs) from the solar panels and losses from inverter and associated systems (K).

Thus, the total irradiance (G) that can be collected from a specific location is:

$$G = \frac{\alpha.\beta.G_i.S_{pv}.N.\zeta_s.K}{1000} \tag{2}$$

To accurately calculate payback periods for solar panels, electricity demand data is acquired to work out the surplus in solar per time of day against usable solar, which will negate the costs of electricity demand to the consumer. Raw data has been extracted from a Journal of Energy Storage [8], which presents UK household electricity demand data for Winter Weekdays, Winter Weekends, Summer Weekdays, and Summer Weekends. The information is credited as data synthesised by the CREST demand model, which used 15,000 households without Solar PV. For the demand we have P_P power produced and can be a positive or negative number, P_{DA} demand and P_{AV} is the available power, which is a function of the average irradiance and system losses [5].

$$P_D = P_{DA} - P_{AV} \tag{3}$$

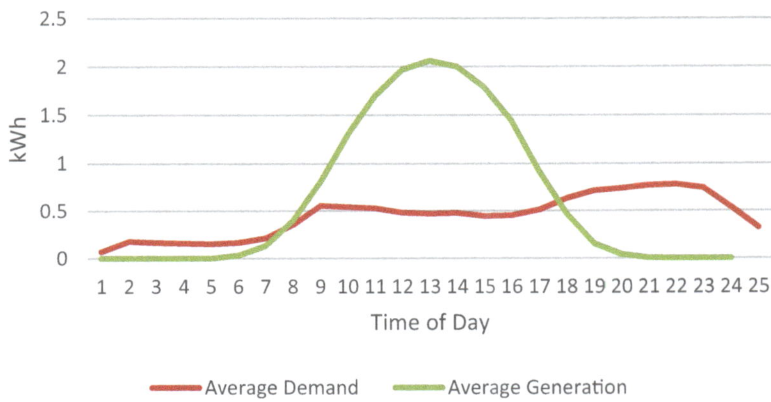

Fig. 2 Simplified annual generation and demand profile

Figure 2 is a simplified annualised graph of generation and demand profile representing the data collected from Eqs. (2) and (3). This clearly shows there is a demand issue during parts of the day with surplus energy being produced in the middle of the day. During the winter months this generation data will fall significantly below the requirements of the household, but it can be shown that during the summer months there is enough irradiance to satisfy the demand of the household and to produce a significant surplus [5].

This provides a number of options to make the most of the solar generated electricity. Option number one is to use the surplus energy to charge a Battery Energy Storage System (BESS) that will discharge energy later in the evening when it is required, thus saving money on electricity usage, and will be the topic of future research. Option number two is to sell surplus energy back to the grid through use of a smart export guarantee (SEG) scheme that exists in the UK and be paid for the surplus energy the solar installation has generated. Using this method there is no need for a battery in the system, but essentially the consumer will be selling their energy at a reduced rate only to have to buy it back at a much higher rate later in the evening. Including a BESS is substantially more cost than solar PV panels alone, but it has the benefit of using the surplus energy to power a household at peak times during the evening, significantly reducing grid dependencies.

For single or multiphase domestic solar generators at a single household who are planning to connect the system to the grid, they must comply with the distribution operation code. In the UK, the technician installing the system must either complete a G98 or G99 form letting the local network aware of the new solar installation so that it can be modelled into the local grid network [9]. The G98 application is for installations of up to 16 Amps, which means a limit on the inverter capacity on the system of 3.68 kW. Whereas a G99 application must be made for larger installations above 16 Amps, for this reason, most installations in the UK are limited to 4 kW.

Summarising the research, as of March of 2023 the best standard electricity tariffs available are supplied by Octopus Energy with a standing charge of 47.52 p/kWh

Table 1 Cost comparison of various scenarios

Scenario	Energy consumption	Cost to consumer in GBP
Original cost of energy	4000 kWh	1550
Cost of energy with solar panels	2200 Wh + generated capacity	921
Cost of energy using solar panels and SEG	2200 Wh + generated capacity + SEG offset	678

and a unit rate of 33.97 p/kWh. If it has been found that a household with average consumption will have an annual bill of around £1550 based on a consumption figure of 4000kWh, this can be reduced to a requirement of 2200 kWh with a solar PV system.

The total cost of consumer demand electricity with the 4 kW solar generation factored in, but not yet including any surplus via the Smart Export Guarantee scheme (SEG) previously known as feed-in tariffs (FIT), giving an annual total of £921, making use only of the solar power that is generated at or below the demand profile for consumer electricity a saving of £614 has been made without taking into account for surplus solar generation that can be sold back to the grid.

The feed-in tariff scheme was closed to new applicants as of 1 April 2019. It has since been relaunched as the Smart Export Guarantee scheme since the 1st of January 2020, which pays small-scale generators for exporting electricity back to the grid. The new SEG scheme works similar of a regular energy supplier where different companies will compete to offer attractive terms and rates, and where if a supplier is no longer attractive the customer can switch to a different SEG licensee for better rates [10].

As of March 2023, the Energy company offering the best SEG was Octopus energy, with a rate of 15 p, only available if you also import with this energy supplier. This equates to a SEG value of £563, thus if we deduct this from our solar bill £921, we are left with an annual bill of £678. Cost comparisons are shown in Table 1 for the three most common scenarios.

3 Grid-Tied Payback Periods

Over the last decade from 2010 to 2021 excluding the rapid rise in energy prices as a result of the conflict in Ukraine, the annual cost of electricity has risen by 38%. This means on average there is a typical increase in electricity prices of 3.45% per year. The price that can be expected to pay for a 4 kW solar installation with no batteries will vary quite drastically between companies and also depend on the quality of hardware used and complexity of the installation. Although the general numbers for a system of this capacity range from £5000 to £8000 [11]. We have shown in Fig. 3, the high-level data for three scenarios of installation costs £5 k, £6.5 k, and £8 k, using a 4 kW soar array, thus we can investigate payback periods.

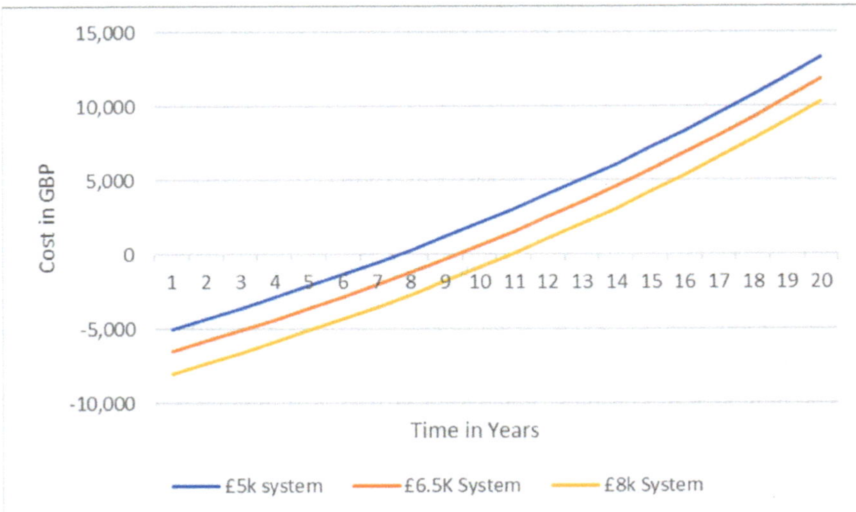

Fig. 3 Payback periods

Adjusted for inflation we can see a payback period of between 8 and 9 years; this would be attractive to a majority of homeowners, who intend to stay in a family home for a number of years. The average length of stay in a UK home is 21 years according to the Office for National Statics [12].

4 Future Research

We are continuing the research to include BESS system into the model and calculate payback periods and break-even points for grid tied systems. We also investigate the use of Solar PV with BESS to establish an off-grid platform for domestic consumers.

5 Conclusion

There is a hesitancy in the UK, to invest in solar energy due to the perceived extended payback periods, and the lack of financial initiatives for working families. This chapter highlights the fact that in the current climate this is a misconception and with the rising cost of energy, it is evident that the adoption of Solar Panels as a means of offsetting the costs of electricity are extremely viable. The projected cost of electricity in 2023 was £678 based on the average percentage increase in electricity prices over the last decade. This means that the projected annual bill of

£678 is just 44% of the actual annual bill of £1535 in 2023 based on a home with an annual demand of 4000 kWh.

The investment into a 4 kW solar installation in 2023 could see annual savings of £614, or up to £1177.54 with the use of a SEG feed-in tariff. This could see payback periods as low as 6 years and at most 11 years depending on the installation costs.

This is made even more attractive to some low-income households with the ECO 4 grant government schemes that can potentially subsidise free solar installations which would see huge benefits to some of the poorest in the community who are struggling with the rising costs of living.

References

1. UK domestic energy consumption trends. www.gov.uk/government/statistics/energy-consumption-in-the-uk-2022
2. UK Energy consumption data Energy Consumption in the UK 2021 (publishing.service.gov.uk). Accessed Dec 2022.
3. Constant, E., Thanapalan, K., & Bowkett, M. (2018). Optimal energy storage evaluation of a solar powered sustainable energy system. In *The proceedings of the 5th IEEE- SICE international symposium on control systems, Tokyo, Japan, March 9–11, 2018.*
4. Constant, E., Thanapalan, K., & Bowkett, M. (2018). *Optimal energy storage evaluation of a solar powered sustainable energy system.* Institute of Electrical and Electronics Engineers.
5. Constant, E., Thanapalan, K., & Bowkett, M. (2019). System sizing for solar powered sustainable energy system. In *International conference on renewable energies and power quality* (pp. 1–5).
6. Cribb, J., Waters, T., Wernham, T., & Xu, X. (2022). *Living standards, poverty and inequality in the UK: 2022.* Institute for Fiscal Studies.
7. European Commision. (2023). *PVGIS online tool.* Retrieved March 2023, from https://joint-research-centre.ec.europa.eu/pvgis-online-tool_en
8. Pimm, J., Cockerill, A. T., & Taylor, G. P. (2018). Time-of-use and time-of-export tariffs for home batteries: Effects on low. *Journal of Energy Storage*, 450.
9. Northern Power Grid. (2023). *Explaining G98 and G99.* Retrieved March 2023, from https://www.northernpowergrid.com/your-powergrid/article/explaining-g98-and-g99
10. GreenMatch. (2023, April 13). *How much does a solar battery storage system cost.* Retrieved from GreenMatch: Ofgem. (2023). *Smart Export Guarantee (SEG).* Retrieved March 2023, from https://www.ofgem.gov.uk/environmental-and-social-schemes/smart-export-guarantee-seg
11. https://www.greenmatch.co.uk/blog/2018/07/solar-battery-storage-system-cost
12. UK Population Data. https://www.ons.gov.uk/peoplepopulationandcommunity/birthsdeathsandmarriages/families/bulletins/familiesandhouseholds/2021
13. Figure 1, Grid connected schematic. https://www.solarreviews.com/blog/grid-tied-off-grid-and-hybrid-solar-systems

Effect of 2023 European Heatwave on Photovoltaic Energy Generation: A Case Study of Central and Southern Italy

Muhammad Ehtsham and Marianna Rotilio

1 Introduction

Understanding how heatwaves affect photovoltaic solar energy systems is crucial for determining the full potential of photovoltaic installations in the context of climate change scenario. Depending on the exact circumstances and the installation's design, heatwaves can affect photovoltaic solar energy systems in both good and negative ways [1]. During heatwaves, sunshine and solar irradiance levels are frequently higher. Because solar panels perform better in brighter, sunnier conditions, photovoltaic systems may generate more electricity as a result [2]. On the other hand, while solar panels profit from more intense sunshine, their efficiency may decline as temperatures rise. The majority of photovoltaic panels perform less effectively at higher temperatures, and their output decreases as the temperature rises [3]. This is due to the fact that as panels heat up, the efficiency of conversion of solar energy to electrical energy decreases. Photovoltaic panels may experience thermal stress in extremely hot conditions, particularly if they are improperly ventilated. The lifespan and long-term performance of the panels may be impacted by this stress, which could eventually cause degeneration or failure [4].

As per Copernicus Climate Change Service, The June–July-August (JJA) season for 2023 was the warmest season globally, with an average temperature of 16.77 °C, 0.66 °C above average. Summer 2023 saw marine heatwaves in several areas around Europe, including around Ireland and the UK in June, and across the Mediterranean in July and August [5]. Global-mean surface air temperatures for the past 18 summers (June-July-August) is depicted in Fig. 1.

M. Ehtsham (✉) · M. Rotilio
Department of Civil, Construction-Architectural and Environmental Engineering, University of L'Aquila, L'Aquila, Italy
e-mail: muhammad.ehtsham@graduate.univaq.it

P. Pong (ed.), *Renewable Energy Resources and Conservation*, Green Energy and Technology, https://doi.org/10.1007/978-3-031-59005-4_5

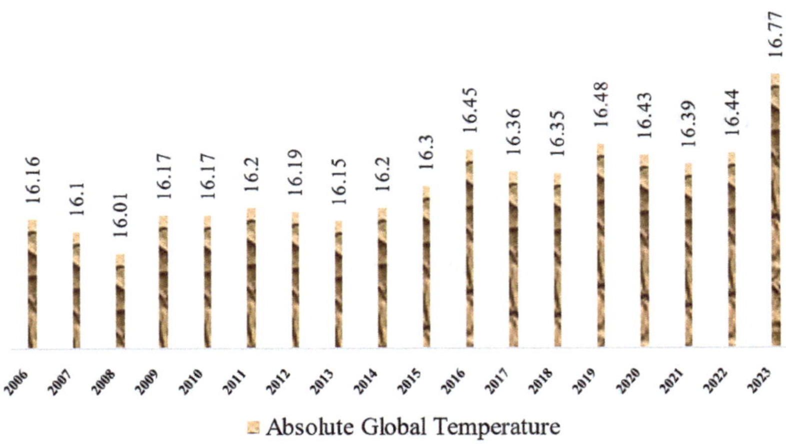

Fig. 1 Global-mean surface air temperatures for the past 18 summers (June-July-August). (Data source: ERA5. Credit: C3S/ECMWF [5])

The objective of this research work is to provide a review of the effect of heatwave on the photovoltaic installations. This research work also assesses how the sudden rise of temperature, which affected the masses across the European and Mediterranean mainland's during the summer 2023, affected the energy generation capacity of the already operational photovoltaic installations. This research work produces some viable results, which may be used by decision makers to take optimal steps during installation and operation of the photovoltaic plants for better operation of plants during the heatwaves.

2 Data Collection

This research study utilized data obtained from the monitoring system of SOLIS.SPA. The monitoring of the photovoltaic plants of the company is carried out using the self-powered self-standing dataloggers attached with the inverters. These dataloggers record the temperature and instantaneous energy at every 5 min and send the records to the central database of the company. A consistent record is kept using the mySQL database, which is accessed by monitoring system of the company and for the research purposes as well. It is noteworthy here to mention that internet connection should be maintained during the operation time to make sure sending and receiving of the data.

Data from 98 inverters associated with three different plants were collected for the period spanning from January 1, 2017, to August 20, 2023. Daily energy production data in kWh was extracted from the mySQL database of the monitoring system using Python code in PyCharm. To establish the connection between Python and the mySQL database, the pandas and mySQL.connector libraries were imported into the Python project's work environment. The database was accessed using specific credentials, and the output file in CSV format was saved at the designated excel_file_path. An in-depth analysis was conducted to assess the continuity of the data. The plant details are presented in Table 1. The study area and approximate locations of the photovoltaic plants are depicted in Fig. 2. All the solar photovoltaic installations are Monocrystalline JA Solar 385 Wp, fixed-axis (FA) arrangements with central inverters (CI). Each inverter is equipped with a built-in maximum power point tracking (MPPT) feature, optimizing energy output to its maximum potential under prevailing atmospheric conditions.

3 Results and Discussion

To assess the performance of plants during heatwave of 2023, performance of plants is compared with performance indicators during previous years. For the ease of visualization of impact of heatwave, the results are depicted during three different time spans; First—1 July to 15 July, Second—16 to 31 July, Third—01 to 20 August. Different indicators are compared with the same three-time spans during past years.

Figure 3 depicts the average solar irradiance and average daily energy generation during three-time spans, i.e., 1–15 July, 16–31 July, and 01–20 August. It is evident from the results that plants A and B generated lower energy in year 2023 as compared to previous years. However, plant C shows a consistent energy generation during all three-time spans for all consecutive years, including 2023.

Different performance metrics, i.e., Reference Yield, Final Yield, Performance Ratio, and Capacity Utilization Factor are calculated as per guidelines of IEC-61724 [6]. A comparison based upon the IEC-61724 is necessary for the performance analysis of photovoltaic plants [7], description of performance metrics is provided in the Table 2. Performance metrics calculated for the selected time spans are depicted in Fig. 4. Plants A and B depicted lower final yield and performance ratio as compared to previous years in almost all three-time spans. Comparatively the performance indicators of plant C are consistent as compared with previous years.

Box plots are created to represent a graphical summary of monthly performance ratio, capacity utilization factor, monthly refence yield, and monthly final yield of each plant during past 7 years. The centerline that divides the two boxes depicts the median value, the middlebox represents the range in which 50% values lie, and the lower and upper whiskers depict the values that are not in the middle 50% range [8]. Figure 5 depicts the box plots of the monthly performance ratio and capacity utilization factor, while Fig. 6 illustrates the monthly reference and final yields during the monitored period, i.e., from January 1, 2017, to August 20, 2023.

It is important to compare the amount of energy generated by each plant during the selected span of the time in each year of monitoring. In Fig. 7, a visualization of the energy generation until August 20 of each year is provided. It is clear that the average daily energy produced in 2023 by the plants A and B is lowest as compared to all the previous years. However, we see a different trend in case of plant C, which

Table 1 Photovoltaic plants specifications

Plant	Type of plant	Location lat.-long.	Rated peak power, KWp	No. of inverters	Range of peak inverter rating	Installation
A	Grid connected	42.11–14.33	347.87	26	10.16–16.23 kWh	Ground mounted fixed axis
B	Grid connected	41.33–13.42	674.67	55	10–12.5 kWh	Rooftop installation fixed axis
C	Grid connected	42.16–14.69	199.7	17	6.72–17.82 kWh	Rooftop installation fixed axis

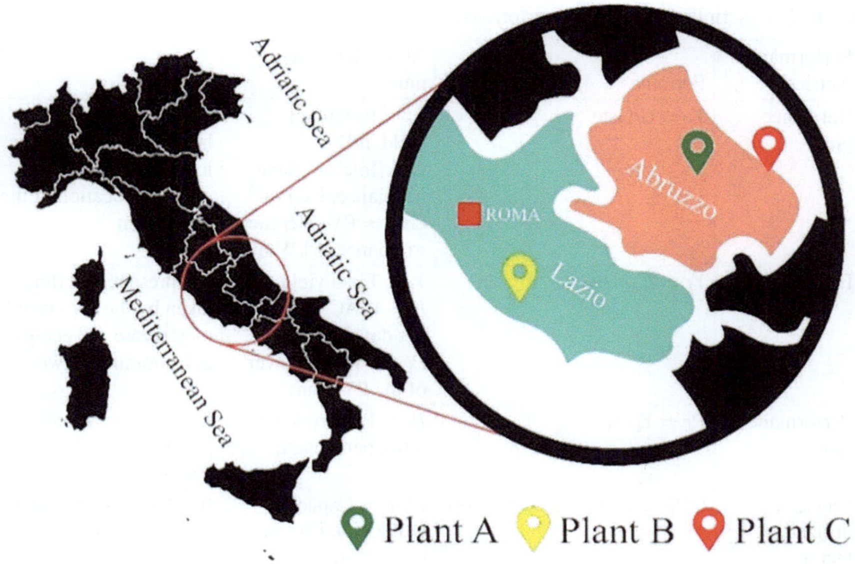

Fig. 2 View of photovoltaic plants installed in the Lazio and Abruzzo regions of Italy

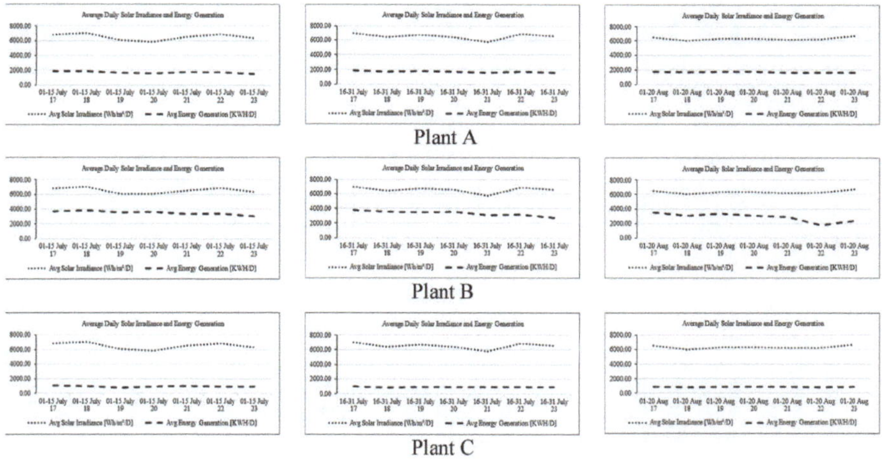

Fig. 3 Comparison of average daily solar irradiance and average daily energy generation during selected time (01–15 July, 16–31 July, and 01–20 August) for the past 7 years of all three plants

shows similar trend with previous years and no anomalies are found in terms of energy generation in 2023.

Table 2 Description of different performance metrics

Performance metrics	Formula	Abbreviation and units	Summary
Reference yield	$Y_r = G/\text{Gstc}$	Y_r = Reference yield, h/D G = Total in-plane irradiance, kWh/m^2 Gstc = PV reference irradiance, 1 kW/m^2	It is the number of hours at the reference irradiance. It depends upon the location of the PV system
Final yield	$Y_f = E_{\text{AC}}/P_{\text{PV}}$	Y_f = Final yield, h/D E_{AC} = AC energy per day, kWh/D P_{PV} = Rated power of the PV plant	It represents the time taken by the PV array to generate AC energy at its nominal power
Performance ratio	$Pr = Y_{\text{F}}/Y_R$	P_r = Performance ratio, percentage	It shows the actual performance of a PV system
Capacity utilization factor	$\text{CUF} = (Y_f/24 \times N_m) \times 100$	CUF = Capacity Utilization Factor, percentage N_m = Number of days	It is the ratio of actual 24-h electricity output from a PV system to its maximum potential yield at rated power

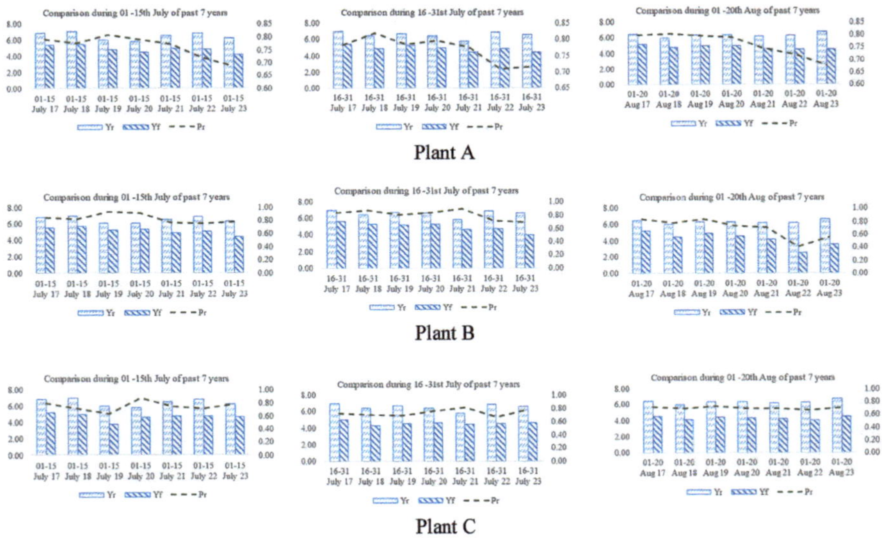

Fig. 4 Comparison of *Yr*, *Yf*, and *Pr* during July–August of past 7 years of all plants

4 Conclusions

This study provides important information about how heatwaves affect the production of photovoltaic energy. Higher radiation, reaching the photovoltaic cells

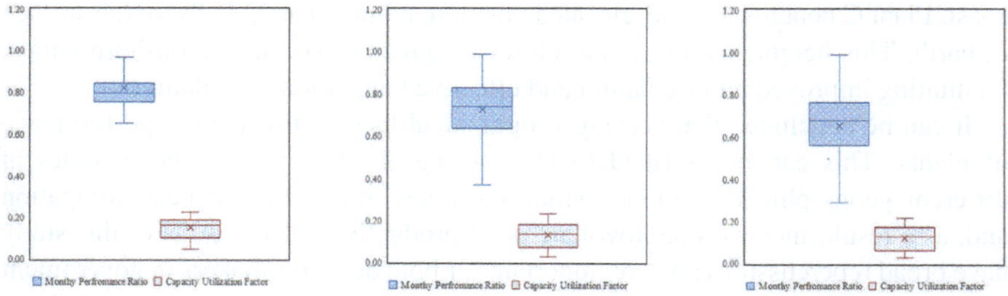

A B C

Fig. 5 Box plots of monthly performance ratio and monthly capacity utilization factor of all three plants during the monitored period

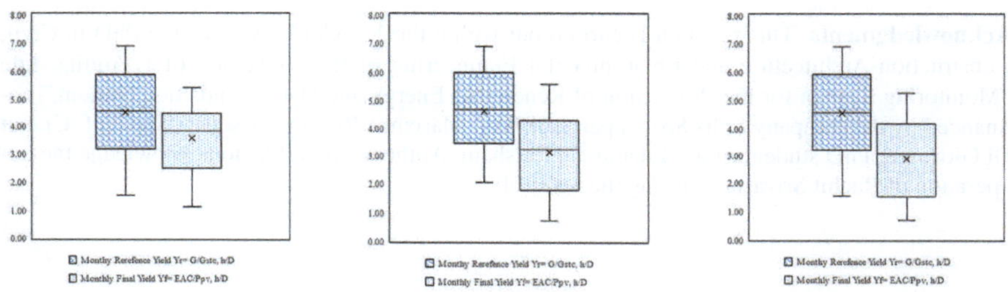

A B C

Fig. 6 Box plots of reference and final yields during the monitored period

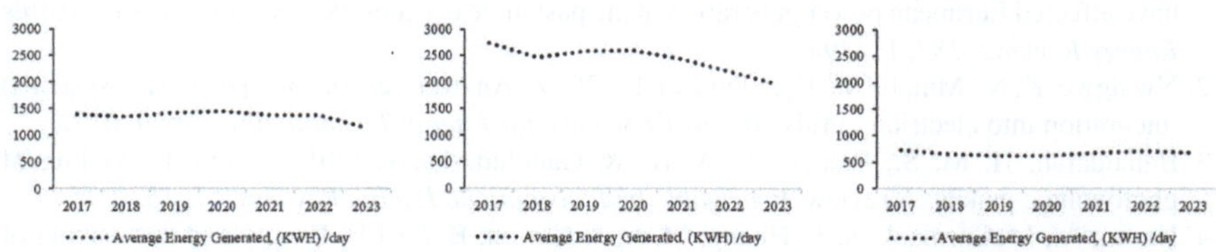

A B C

Fig. 7 Average energy generation per day in kWh by each plant in past 7 years

improves their productivity, it also raises the module temperature, which has a negative impact on the efficiency of the entire plant [9]. With the increase of temperature of photovoltaic modules, the electrical efficiency decreases because photovoltaic modules convert 20 percent of solar energy into electricity and rest 80 percent is dissipated as heat losses [10]. Three plants were analyzed based upon IEC-61724 guidelines and results revealed a noteworthy trend: Plant A, located in a mountainous area, saw a considerable decrease in energy output due to the extreme heatwave in the summer of 2023. Plant B, which is situated almost 10 km from the Mediterranean coast, also had a similar pattern. However, the sustained presence of onshore winds during the day has played a pivotal role in maintaining Plant C's consistent performance relative to previous years. Situated along the Adriatic coast, Plant C benefits from its elevated position, mounted at 22 ft above the surface of earth. This heightened elevation allows for greater exposure to onshore winds, facilitating improved air circulation and effective temperature regulation.

It can be concluded that varying geographical factors affected the performance of plants. This can be ascribed to the varying air flow around the modules at different geographical positions, which promotes more efficient heat dissipation and, as a result, increases photovoltaic plant productivity. The results of this study have broad repercussions and are important for both decision-makers in government and businesses engaged in the design and construction of photovoltaic installations. Stakeholders can optimize the performance and durability of solar plants in the face of rising temperatures and more frequent heatwaves by taking these findings into account.

Acknowledgments This research is carried out within the XXXVIII Cycle of the PhD in Civil, Construction-Architecture and Environmental Engineering of the University of L'Aquila, title "Monitoring System for the Production of Renewable Energy and Data Acquisition System," co-financed by the company Solis Spa, supervisor Prof. Marianna Rotilio, co-supervisor Prof. Gianni Di Giovanni, PhD student Eng. Muhammad Ehtsham. Authors would like to acknowledge the co-operation of Rachit Srivastava during the research.

References

1. Brás, T. A., Simoes, S. G., Amorim, F., & Fortes, P. (2023). How much extreme weather events have affected European power generation in the past three decades? *Renewable and Sustainable Energy Reviews, 183*, 113494.
2. Nwaigwe, K. N., Mutabilwa, P., & Dintwa, E. (2019). An overview of solar power (PV systems) integration into electricity grids. *Materials Science for Energy Technologies, 2*, 629–633.
3. Bahaidarah, H. M. S., Baloch, A. A. B., & Gandhidasan, P. (2016). Uniform cooling of photovoltaic panels: A review. *Renewable and Sustainable Energy Reviews, 57*, 1520–1544.
4. Hasan, A., McCormack, S. J., Huang, M. J., & Norton, B. (2014). Energy and cost saving of a photovoltaic-phase change materials (PV-PCM) system through temperature regulation and performance enhancement of photovoltaics. *Energies (Basel), 7*, 1318–1331.
5. Copernicus: European Union Earth Observation Programme. https://climate.copernicus.eu/surface-air-temperature-maps. Accessed 02 Sept 2022.
6. Klise, K. A., Stein, J. S., & Cunningham, J. (2017). Application of IEC 61724 standards to analyze PV system performance in different climates. In *2017 IEEE 44th photovoltaic specialist conference (PVSC)* (pp. 3161–3166). IEEE.
7. Srivastava, R., Tiwari, A. N., & Giri, V. K. (2020). An overview on performance of PV plants commissioned at different places in the world. *Energy for Sustainable Development, 54*, 51–59.
8. Muhammad, E., Muhammad, W., Ahmad, I., Muhammad Khan, N., & Chen, S. (2020). Satellite precipitation product: Applicability and accuracy evaluation in diverse region. *Science China Technological Sciences, 63*, 819–828.
9. Hasan, K., Yousuf, S. B., Tushar, M. S. H. K., Das, B. K., Das, P., & Islam, M. S. (2022). Effects of different environmental and operational factors on the PV performance: A comprehensive review. *Energy Science & Engineering, 10*, 656–675.
10. Rahman, M. M., Hasanuzzaman, M., & Abd Rahim, N. (2017). Effects of operational conditions on the energy efficiency of photovoltaic modules operating in Malaysia. *Journal of Cleaner Production, 143*, 912–924.

Classification of Types of Daily Solar Radiation Patterns Using Machine Learning Techniques

Sebastián Alvarez-Flores, Kevin Guamán-Charro, Enrique Yupa-Loja, and Xavier Serrano-Guerrero

1 First Section

1.1 Introduction

The global energy demand has been steadily increasing in recent decades due to population growth, industrial development, and rising per capita consumption. However, the generation of energy from fossil fuels has led to a significant increase in greenhouse gas emissions, resulting in global temperature rise and unprecedented climate change. Therefore, it is crucial to seek renewable and sustainable energy sources to meet global energy demand.

According to the International Renewable Energy Agency (IRENA) report, renewable energy could provide up to 90% of global electricity by 2050, significantly reducing greenhouse gas emissions [1]. Furthermore, renewable energy can offer a more cost-effective and long-term sustainable energy source, as demonstrated in various studies [2, 3].

The methodology employed by most studies [4] to develop a solar irradiation classification model begins with the definition of parameters related to climate stochasticity. Based on these different input parameters, three existing model types are defined. The first is known as the temporal factors model, which utilizes different constant weather conditions such as solar angle, day length, local time, etc. Consequently, these models can only be applied to calculate long-term averages. The second is the Gaussian functions model, which assumes that every daily solar pattern approximates a Gaussian trend, but this hypothesis is inconclusive for unclear days. Finally, the direct function model defines the relationship between

S. Alvarez-Flores (✉) · K. Guamán-Charro · E. Yupa-Loja · X. Serrano-Guerrero
Energy Transition Research Group, Universidad Politécnica Salesiana, Cuenca, Ecuador
e-mail: malvarezf@est.ups.edu.ec

P. Pong (ed.), *Renewable Energy Resources and Conservation*, Green Energy and Technology, https://doi.org/10.1007/978-3-031-59005-4_6

solar radiation and various meteorological parameters, where the parameters are linked in a function. However, parameters like the clarity index lend themselves to generalizations and do not demonstrate their intrinsic relation to climatic conditions. Thus, a cloudy day does not necessarily mean a less clear day.

This chapter describes the identification of solar radiation patterns in the city of Cuenca, Ecuador, using real data measured between 2014 and 2017. This methodology is approached through conditioning, classification, and recognition techniques that lead to the determination of different types of solar radiation through a novel model based on converting temporal series data into graphical representations for classification through neural networks [5]. The most significant contributions of this work are described below.

Complementary Model The obtained results for a specific temporal pattern of solar irradiation can be compared with other studies that use assisted methodologies as previously described (related to climate stochasticity). In this way, applying, for example, the direct function model, some of the patterns identified by the neural network are demonstrated punctually, such as characteristics of cloudiness, rainfall, summer solstices, etc.

Generalized Model By experimenting with unassisted classification models, there is an advantage in not requiring extensive knowledge of the geography or known climatic variations of a specific region. Consequently, the classification results will only clearly indicate irradiation variations, making it useful in energy efficiency applications and applicable in different parts of the world.

Previous studies have placed emphasis on improving energy efficiency through the application of various methodologies. For instance, in [6], a model is defined that proposes the integration of Distributed Resources (DR) in buildings. It is mentioned that one potentiality of this model is the ability to conduct annual energy balance analyses to anticipate future implementations of an energy system. In [7], it is concluded that analyzing the effects of orientation and irradiance would be highly beneficial. Furthermore, in [8], the authors define a model that encompasses different electrical consumption profiles as a strategy to enhance electrical systems. These analyzed studies, along with others [9, 10], in conjunction with the present chapter, can result in the implementation of more efficient systems.

The remaining chapter is structured as follows: Section II provides the necessary background to understand the manuscript's content. Section III describes the step-by-step methodology as indicated in Fig. 2. Section IV presents the results and discussion regarding the results of other authors who have employed different classification methodologies. Finally, Sections V and VI present the conclusion and bibliography, respectively.

1.2 Background

Angular Gramian Field (AGF) The Angular Gramian Field (AGF) represents time series data in a polar system, where each element is the cosine of the sum of angles. Given a time series $X = \{x1, x2, \ldots, xn\}$, where n is a real number, the values are scaled to be within the range of [−1,1] or [0,1] using Eq. 1:

$$x_{-1}^{\sim i} = \frac{\left(xi - \max(X) + (xi - \min(x))\right)}{\max(x) - \min(x)} \tag{1}$$

Here $x_{-1}^{\sim i}$ is the i-th normalized value in the dataset, xi is the i-th value of the initial dataset, min(x) is the desired minimum value, and max(x) is the desired maximum value.

In this polar coordinate system, the rescaled values of the time series are represented as follows:

$$\begin{cases} \varnothing = \arccos\left(\tilde{x}_i\right), -1 \leq \tilde{x}_i \leq 1, \tilde{x}_i \in \tilde{X} \\ r = \frac{t_i}{N}, t_i \in \mathbb{N} \end{cases} \tag{2}$$

$\tilde{X}$ represents the set of rescaled data. t_i is the time mark, and N is a constant factor that regulates the space of the polar coordinate system. This system allows for visualizing and understanding data related to time series from a different perspective.

Convolutional Neural Networks (CNNs) The main idea behind Convolutional Neural Networks (CNNs) is to obtain features through a hierarchical methodology. This means that features such as color and contour are obtained from the early layers, while middle filters extract textures constructed from the composition of colors and edges. Finally, the last filters extract specific textures and patterns from each image. Therefore, higher-level features are built upon lower-level features, and so on. In this way, CNNs are capable of robustly extracting features.

K-Means The k-means clustering technique relies on a value of k, which must be determined prior to performing any clustering analysis. Clustering with different values of k will eventually yield different results. Several techniques can be used to determine the value of k, including the Calinski-Harabasz Index and Davies Boulding.

Davies Boulding Index Is a metric for clustering algorithms evaluation, the Davies-Bouldin index determines how cohesive a cluster is. It is defined as follows:

$$DB = \frac{1}{k} \sum_{i=1, i \neq j}^{k} \max \left(\frac{\sigma_i + \sigma_j}{d\left(c_i, c_j\right)} \right) \tag{3}$$

where k is the number of clusters, σ_i is the mean distance between each point in the cluster i and the cluster centroid, σ_j is the mean distance between each point of j cluster. Finally, $d(c_i, c_j)$ is the distance between bought centroids.

2 Second Section

2.1 Methodology

The methodology, as shown in Fig. 1, begins with exploratory analysis of the initial data, which corresponds to the solar irradiation time series. In this analysis, information is extracted, processed, and a portion of it is visualized to observe some of the daily patterns in a general way. Then, an initial matrix is established, which will be useful for the subsequent stages. The next part refers to the transformation of the temporal data into polar image representations using the "Angular Gramian Field" (AGF). In the next step, the images are input into a CNN with VGG16 architecture. Finally, with the help of clustering methodologies, the information can be differentiated into k clusters.

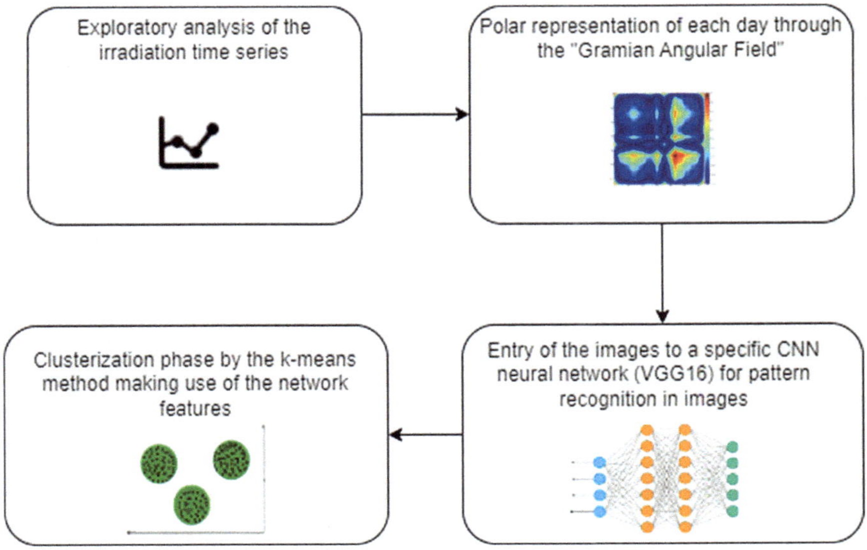

Fig. 1 Description of the methodology used for the classification of solar patterns

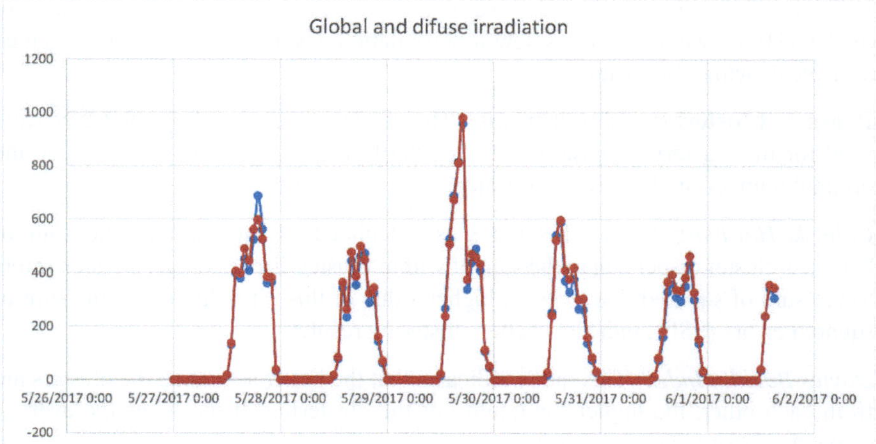

Fig. 2 Global and diffuse irradiation in Cuenca Ecuador between May 27 and June 1, 2017

Fig. 3 Polar representation of different days

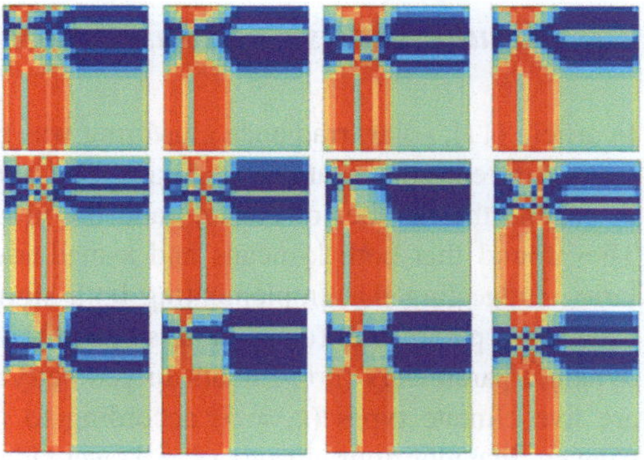

Phase 1: Exploratory Data Analysis The first phase of the methodology involves conducting an exploratory data analysis. Daily global and diffuse irradiation data were collected over a period of 4 years. Fig. 2 displays the plot of the global irradiation in red and the diffuse irradiation in blue, covering the period from May 7 to June 1, 2017.Since the difference between the pair of irradiation data is practically negligible, for the purposes of this study, global irradiation was used as a reference for the remaining stages.

Phase 2: Polar Representation Through the Angular Gramian Field As shown in Fig. 3, the data is separated by day and encoded into their polar representation using the Angular Gramian Field (AGF) transform. This results in a total set of 1194 images.

Phase 3: Application of Convolutional Neural Network (CNN VGG16) For pattern recognition, the obtained images were fed into a pretrained CNN, specifically the VGG16 model, which can recognize specific patterns in images. The resulting vector representation for each day was extracted from the nodes of the last layer of the VGG16 network. With this new representation, each daily series is represented as a vector with 4095 features.

Phase 4: Clustering Using k-means The attributes generated by the CNN were used for the clustering phase, employing the k-means algorithm to determine the optimal number of clusters. Two strategies were explored.

Calinski-Harabasz Index This index is calculated as the ratio of the sum of between-cluster dispersion to the sum of within-cluster dispersion. Dispersion refers to the sum of squared distances. A higher value of this ratio, known as the sum of squared errors (SSE), indicates better clustering results.

Davies Bouldin Index This index indicates that the better separated the clusters are from each other, the higher the quality of the clustering. It measures the average dispersion between each cluster.

2.2 Results Analysis and Discussion

In article [11], solar irradiance clustering was performed based on time series data from different solar stations in Spain. The authors compared two values of k to determine the number of clusters and the most efficient clustering methodology. They found that both k-means and k-medoids were efficient, with a 90% similarity, suggesting that implementing both simultaneously was unnecessary. They used static parameters (dispersion, symmetry, and distribution characteristics) and dynamic parameters (non-Euclidean geometry) to extract features. In Spain, there are five climate types ($k = 5$) according to the Koppen classification, while in Cuenca, Ecuador, four clusters ($k = 4$) were identified as optimal.

In article [11], k-means clustering is used to evaluate photovoltaic systems over a 5-year period in Malaysia. The study identifies optimal values for k as 3, 4, and 5, with 4 being chosen. The clusters are categorized as overcast, moderate, mixed, and high variability. Despite the study's location difference from a tropical climate, there is similarity in the number of clusters.

Optimal *k* Determination The decision to apply k clusters using this method can be observed in Figs. 4 and 5. Applying the Davies Boulding criterion does not provide a definitive result for k, as the average minimum distance value is practically 0 after 6 clusters, which would result in overly dense clusters. Additionally, there is no change in trend for earlier values that could indicate an optimal value.

Finally, applying the Calinski-Harabasz criterion reveals an elbow or change in trend at a value of 4. This indicates that higher values would result in lower quality clustering. The goal of applying these criteria is to obtain the maximum number of clusters while maintaining their quality. Therefore, $k = 4$ was chosen as the optimal value.

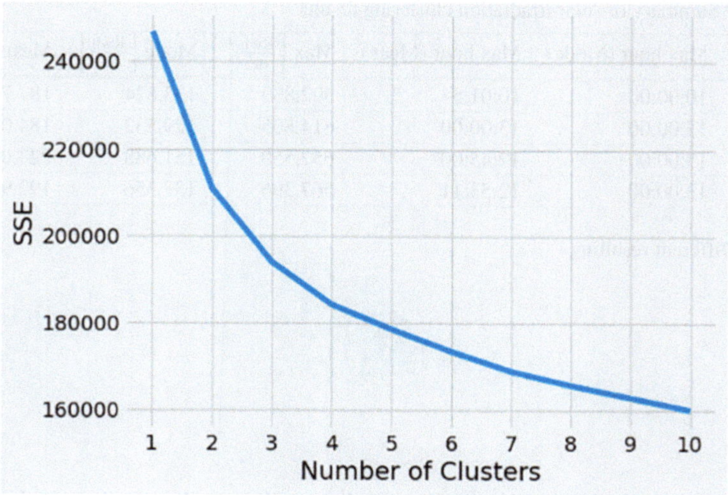

Fig. 4 Calinski-Harabasz index

Fig. 5 Davies Boulding index

Cluster Obtaining Table 1 presents the characteristics of the obtained clusters, including the maximum value, average value, and variance of each dataset. Some notable differences can be observed, where the irradiation varies significantly as shown in Figs. 6 and 7.

Occurrence in Each Month of the Year The frequency of each cluster in each month is depicted in Figs. 8, 9, 10 and 11. Cluster 0 and Cluster 3 are the least frequent, while Cluster 1 and Cluster 2 exhibit higher prevalence throughout the months.

Table 1 Summary of solar irradiation clustering results

Cluster	Max hour (Mode)	Max hour (Mean)	Max $\left[\frac{Kwh}{m^2}\right]$	Mode $\left[\frac{Kwh}{m^2}\right]$	Mean $\left[\frac{Kwh}{m^2}\right]$
0	10:00:00	10:01:54	592.889	133.824	187.715
1	13:00:00	13:00:00	614.805	129.532	184.092
2	13:00:00	12:45:00	657.550	151.688	223.087
3	13:00:00	12:53:11	567.396	137.356	192.935

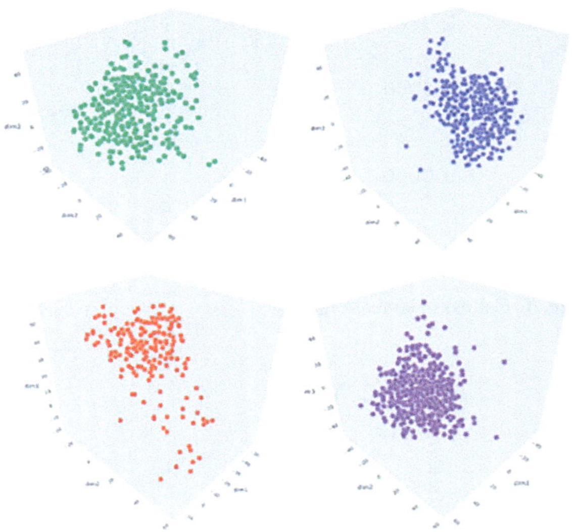

Fig. 6 Different resulting clusters

Furthermore, during the observed period, a significant decrease in the frequency of Cluster 0 can be observed between each year. This cluster, characterized by higher irradiation levels in the morning and lower levels in the afternoon, may be influenced by seasonal variations and environmental changes over time.

Classification Results and Adaptability While the clusters resulting from neural network classification may not provide clear information about specific climatic parameters or other factors, they offer a generalized understanding of possible seasonal patterns. This brings several advantages when applying this methodology to different geographic areas, allowing for the abstraction of inherent climate stochasticity.

In addition to the criteria used, such as the Davies Boulding and Calinski-Harabasz indices, the conclusions of several mentioned authors in the study were considered. These authors determined that solar radiation can be classified into three to four types, including clear sky, intermittently clear sky, completely cloudy sky, and intermittently cloudy sky. The author of this study [4] suggests that most existing research concludes that solar irradiation can be classified into four types, and regardless of location and angle of incidence, only 4 years of daily solar irradiation samples are needed to classify the data into three typical daily classes.

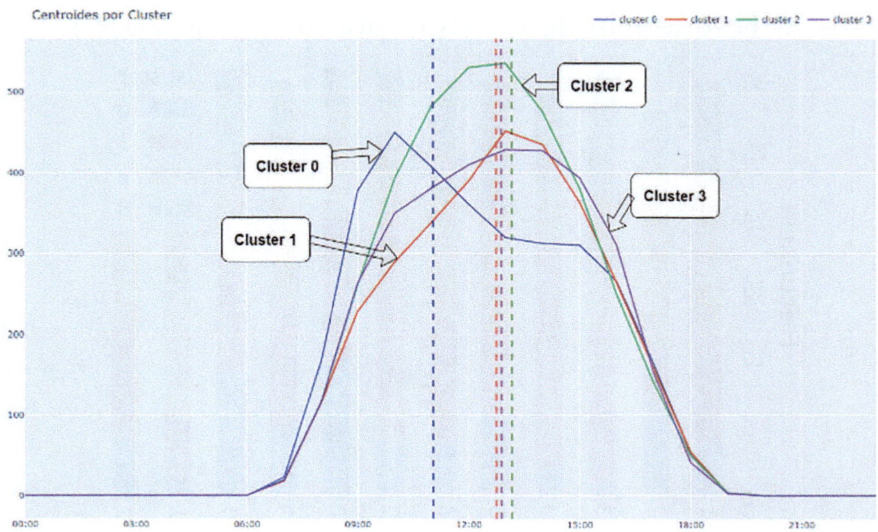

Fig. 7 Clusters represented in the form of daily pattern models

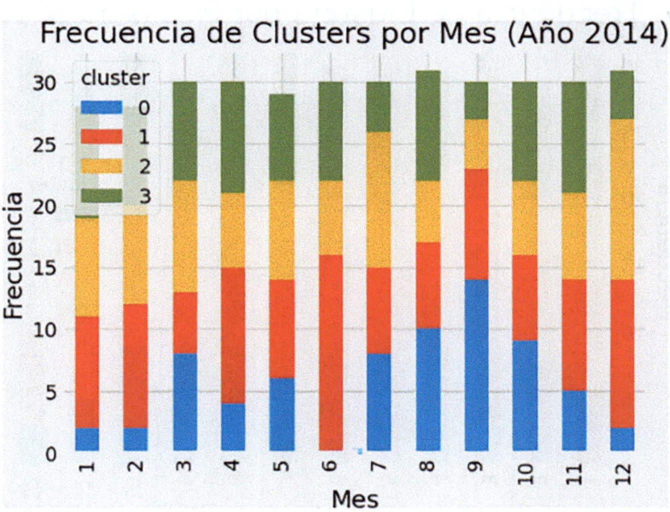

Fig. 8 Occurrence of clusters in 2014

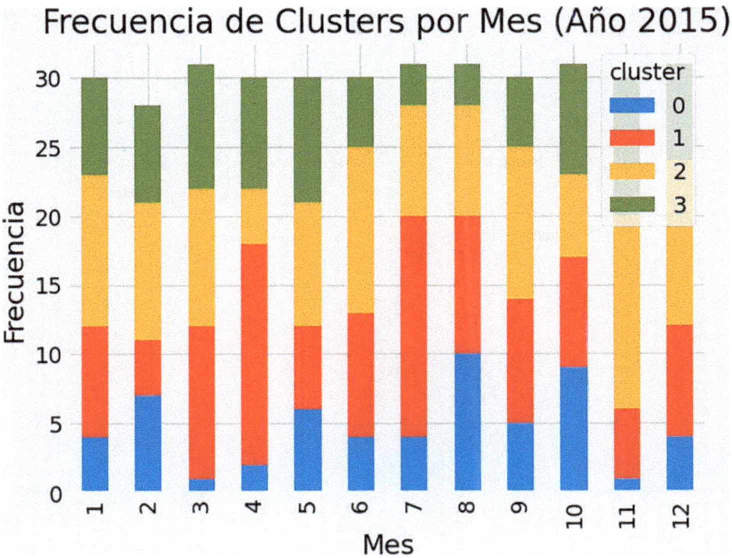

Fig. 9 Occurrence of clusters in 2015

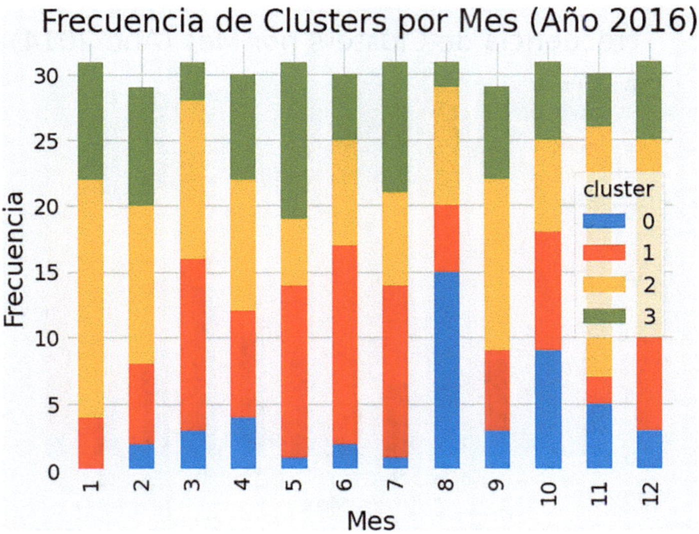

Fig. 10 Occurrence of clusters in 2016

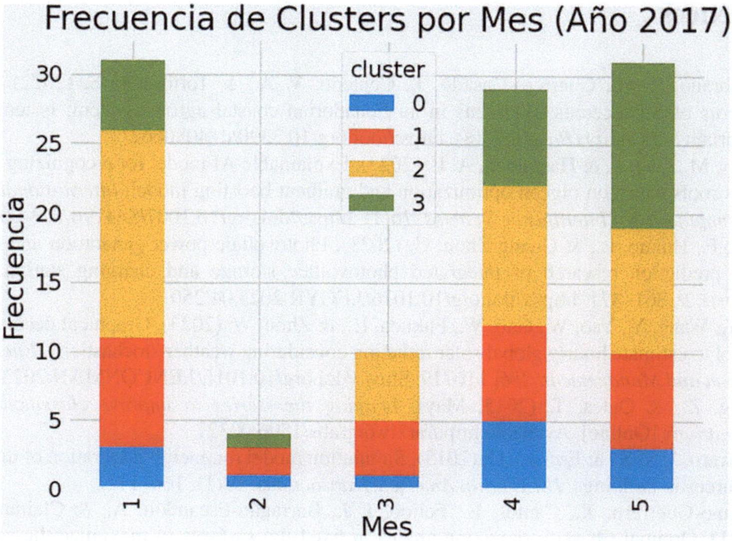

Fig. 11 Occurrence of clusters in 2017

The data used in this experiment were collected in the city of Cuenca, Ecuador, from January 3, 2014, to May 31, 2017.

3 Conclusions

The use of different methodologies in conjunction with machine learning demonstrates in the present study that it is a reliable method that resolves the challenge of manually characterizing climatic parameters. The Davies Boulding and Calinski-Harabasz indices were the initial criteria employed to determine the number of clusters, and their reliability was confirmed by comparing them with results from different studies.

The development of this methodology generates new and more efficient approaches for future studies related to the improvement of solar energy resources. Furthermore, this study can lead to the creation of enhanced automated energy systems tailored to the local climate conditions. In this specific case, applying these advancements to the southern region of Ecuador could result in increased energy production capacity and an improved quality of life for its citizens.

References

1. Zambrano, R. H., Guerrero-Casado, J., Centeno, V. A., & Tortosa, F. S. (2022). Activity patterns of Stenocercus iridescens in an Ecuadorian coastal agroecosystem: Is temperature important? *Diversity (Basel), 14*(8). https://doi.org/10.3390/d14080662
2. Torky, M., Gad, I., & Hassanien, A. E. (2023). Explainable AI model for recognizing financial crisis roots based on pigeon optimization and gradient boosting model. *International Journal of Computational Intelligence Systems, 16*(1). https://doi.org/10.1007/S44196-023-00222-9
3. Tian, F., Huang, L., & Guang Zhou, C. (2023). Photovoltaic power generation and charging load prediction research of integrated photovoltaic storage and charging station. *Energy Reports, 9*, 861–871. https://doi.org/10.1016/J.EGYR.2023.04.250
4. Li, Y., Wang, Y., Yao, W., Gao, W., Fukuda, H., & Zhou, W. (2023). Graphical decomposition model to estimate hourly global solar radiation considering weather stochasticity. *Energy Conversion and Management, 286*, 116719. https://doi.org/10.1016/J.ENCONMAN.2023.116719
5. Wang, Z., & Oates, T. (2015, May). *Imaging time-series to improve classification and imputationy* [Online]. Available http://arxiv.org/abs/1506.00327
6. Guerrero, J. X. S., & Escrivá, G. (2015). Simulation model for energy integration of distributed resources in buildings. *IEEE Latin America Transactions, 13*(1), 166–171.
7. Serrano-Guerrero, X., Cantos, E., Feijoo, J. J., Barragán-Escandón, A., & Clairand, J. M. (2021). Optimal tilt and orientation angles in fixed flat surfaces to maximize the capture of solar insolation: A case study in Ecuador. *Applied Sciences, 11*(10), 4546.
8. Serrano-Guerrero, X., Escrivá-Escrivá, G., Luna-Romero, S., & Clairand, J. M. (2020). A time-series treatment method to obtain electrical consumption patterns for anomalies detection improvement in electrical consumption profiles. *Energies, 13*(5), 1046.
9. Serrano Guerrero, J. X. (2020). *Caracterización de la demanda de energía mediante patrones estocásticos en las Redes Eléctricas Inteligentes*. Doctoral dissertation, Universitat Politècnica de València.
10. Milton, M. A., Pedro, C. O., Xavier, S. G., & Guillermo, E. E. (2018). Characterization and classification of daily electricity consumption profiles: Shape factors and k-means clustering technique. In *E3S web of conferences* (Vol. 64, p. 08004). EDP Sciences.
11. Garcia-Gutierrez, L., Voyant, C., Notton, G., & Almorox, J. (2022). Evaluation and comparison of spatial clustering for solar irradiance time series. *Applied Sciences (Switzerland), 12*(17). https://doi.org/10.3390/app12178529

Optimizing Energy Savings in Polyisoprene Production Through Solar-Based Thermal Technology

Ivana Špelić and Alka Mihelić-Bogdanić

1 Introduction

Large-scale industrial processes, especially polymer producing industry, generate large amounts of waste heat, which can be returned to the process and used as a valuable energy source. The sources of waste heat include both heat loss from products, equipment, and processes as well as combustion heat discharge [1]. The heat recovery utilization from waste heat can lead to energy optimization, cost reduction, and a decrease in the environmental pollution. In the last decades, significant concerns were raised regarding greenhouse emission through fossil fuel utilization. Although concerns are raised, the majority of industrial applications still rely on classical fossil fuel sources. Most of the scientist have attempted to facilitate renewable energy options to speed up the clean energy transition and reduce production cost. The world's most abundant permanent source of energy today is definitely solar energy, whose implementation depends on yearly solar radiation and requires backup energy source to ensure full potential on days with less solar radiation [2, 3]. Regarding positive environmental impacts as well as the best energy savings, the combination of solar generated steam and flue gases heat recovery shows the greatest fuel and exhaust flue gases temperature reduction [4, 5]. The utilization of return condensate combined with the flue gases heat recovery for feed water and combustion air preheating is analyzed. The proposed plant includes both an economizer, air preheater, and boiler feed tank combined with solar-based thermal technology. The main novelty presented in this study are extensive energy efficiency improvements in polyisoprene production when applying hybrid system combining solar energy together with flue gases heat

I. Špelić (✉) · A. Mihelić-Bogdanić
Faculty of Textile Technology, University of Zagreb, Zagreb, Croatia
e-mail: ivana.spelic@ttf.unizg.hr

P. Pong (ed.), *Renewable Energy Resources and Conservation*, Green Energy and Technology, https://doi.org/10.1007/978-3-031-59005-4_7

recovery. The environmental analysis showed reduction of natural gas consumption while simultaneously diminishing the flue gases exhaust temperature and volume.

2 Process Data and Methods of the Energy Efficiency Improvements

The purpose of the chapter is to provide an assessment of the quality and quantity of useful waste heat potential in the proposed polyisoprene production (latitude $\varphi = 45°49'$ N, longitude $\lambda = 15°58'$ E) through the two major waste streams. Heat recovery application involves applying both flue gases and condensate heat recovery together with solar energy (ECO + AP + CHR + SOLAR). The results of the study are shown as potential fuel savings. First case study presents the energy consumption in the continuous manufacturing process of polyisoprene production. The basic process needs electrical energy in the amount of $e_e = 0.893$ kWh$_e$/kg$_P$, which is supplied from the grid and used by polymerization vessel ($e_e = 0.261$ kWh$_e$/kg$_P$), dewatering machine ($e_e = 0.328$ kWh$_e$/kg$_P$), baler ($e_e = 0.152$ kWh$_e$/kg$_P$), and packaging machines ($e_e = 0.152$ kWh$_e$/kg$_P$) [6]. The dry saturated steam in the amount of $d_S = 1.320$ kg$_S$/kg$_P$ is produced in a boiler with efficiency $\eta_B = 70\%$ and is supplied to the solvent stripper with a temperature of $t_S = 121$ ° C (394 K). The process also uses cold water for polymerization in the amount of $d_W = 3$ kg$_W$/kg$_P$. The present study of energy consumption in the polyisoprene production process, working in two shifts at rate $D_{P_y} =$ 12000 t$_P$/yearly, is carried out. The plant works 16 h in a day, 25 days in a month ($\tau = 16$ h/day = 400 h/month = 4000 h/year = 25 days/month = 250 days/year = 10 months/year); therefore, the plant use factor becomes $\beta = 45.66\%$ [7].

The production capacity is calculated as $D_P = 3000$ kg/h $= 12 \times 10^6$ kg/year. Feed water enters the boiler with temperature $t_{\text{FW}_\text{B}} = 24°\text{C}$ (297.15 K), and the whole condensate with temperature $t_C = 104.4$ ° C (377.55 K) is withdrawn to the surrounding. The air required for combustion passes into a firebox with temperature $t_{\text{a}_\text{B}} = 24°\text{C}$ (297.15 K), while the temperature of the exhaust stacks is $t_{\text{FG}_{\text{AP}_\text{o}}} =$ 204°C (477.15 K) [6]. The natural gas with composition 0.85 % CO_2, 0.56 % N_2, 98.05 % CH_4, 0.36 % C_2H_6, 0.12 % C_3H_8, 0.05 % C_4H_{10}, and 0.01 % C_5H_{12} is burned with the excess air coefficient $\alpha = 1.25$. Based on this composition, the lower heating value of the fuel is calculated by the following formula [8] as $H_L = 35,516.25$ kJ/m^3.

The heat transferred to the boiler per unit of product is $q_{S_B} = 3442.03$ kJ/kg$_P$ where $h_S(t_S = 121$ ° C (394.15 K)) and $h_B\left(t_{\text{FW}_\text{B}} = 24°\text{C}\right.$ (297.15 K)) are the steam and water enthalpies taken from thermodynamic tables. The dry saturated steam in the amount of $d_S = 1.320$ kg$_S$/kg$_P$ is produced in a boiler. From this data, the unit volume of the fuel requirement using heat balance is $v_{\text{F}_\text{P}} =$ 0.1384 m^3$_\text{F}$/kg$_\text{P}$.

The fuel consumption becomes $V_F = 415.2\ \mathrm{m^3{}_F/h} = 6643.2\ \mathrm{m^3{}_F/day} = 0.166 \times 10^6\ \mathrm{m^3{}_F/month} = 1.66 \times 10^6\ \mathrm{m^3{}_F/year}$. The heat transferred to the boiler becomes $Q_B = 10.32 \times 10^6$ kJ/h $= 165.2 \times 10^6$ kJ/day $= 4.13 \times 10^9$ kJ/month $= 41.3 \times 10^9$ kJ/year.

The specific steam consumption is $d_{S_P} = 9.537\ \mathrm{kg_S/kg_F}$. The heat of the condensate is calculated as $q_{C_B} = 666.6\ \mathrm{kJ_C/kg_P}$, where $h_{104.4\,^\circ\mathrm{C}} = h_C$ is enthalpy value taken from thermodynamic tables. The whole quantity of condensate from the process is calculated as $D_C = D_S = 3960\ \mathrm{kg_C/h}$ with temperature $t_C = 104.4\,^\circ$C (377.55 K) withdrawn to the surrounding.

2.1 Exhaust Product Analysis

The products of fuel combustion are mostly gaseous. For complete gas combustion, 25% excess air is supplied (excess air coefficient $\alpha = 1.25$). The minimum oxygen volume $V^m_{O_2}$ and stoichiometric air volume V_a are required for combustion [9].

The minimum oxygen volume $V^m_{O_2}$ ($\mathrm{m^3{}_{O_2}/m^3{}_{fuel}}$) is calculated as:

$$V^m_{O_2} = 0.01 \times \left[2 \times \mathrm{CH_4} + \sum \left(x + \frac{y}{4}\right) \times \mathrm{C}_x\mathrm{H}_y\right] = 1.9837\ \mathrm{m^3{}_{O_2}/m^3{}_F}$$

The minimum air volume V_a ($\mathrm{m^3{}_a/m^3{}_{fuel}}$) is calculated as $V_a = 9.446\ \mathrm{m^3{}_a/m^3{}_F}$. The actual volume of air calculated with the excess air coefficient $\alpha = 1.25$ is $V_{a\alpha} = 11.807\ \mathrm{m^3{}_a/m^3{}_F}$.

Volume of each gas component is calculated using following expressions:

$$\text{The volume of carbon dioxide } V_{CO_2} = 0.01 \times \left[\mathrm{CO_2} + \sum x\mathrm{C}_x\mathrm{H}_y\right] = 1.0023\ \mathrm{m^3{}_{CO_2}/m^3{}_F}.$$

$$\text{The volume of water vapor } V_{H_2O} = 0.01 \times \left(0.05y \times \mathrm{C}_x\mathrm{H}_y\right) + \left[\left(V_a \times \alpha \times d\right)/\rho\right] = 2.152\ \mathrm{m^3{}_{H_2O}/m^3{}_F},$$

where $d = 0.13\ \mathrm{kg/m^3}$ is air moisture and $\rho = 0.805\ \mathrm{kg/m^3}$ is steam density. The volume of nitrogen $V_{N_2} = \alpha \times [(0.79 \times V_a) + (N/100)] = 9.334\ \mathrm{m^3{}_{N_2}/m^3{}_F}$.

The volume of oxygen $V_{O_2} = 0.21 \times (\alpha - 1) \times V_a = 0.4959\ \mathrm{m^3{}_{O_2}/m^3{}_F}$. The total volume of flue gases is summed by adding together the volume of carbon dioxide, water vapor, nitrogen unconsumed during the combustion process, and excess oxygen $V_{FG_F} = V_{O_2} + V_{H_2O} + V_{N_2} + V_{O_2} = 12.984\ \mathrm{m^3{}_{FG}/m^3{}_F}$.

The component percentage in flue gases' composition are $V_{CO_2} = 7.72\%$, $V_{H_2O} = 16.57\%$, $V_{N_2} = 71.89\%$, $V_{O_2} = 3.82\%$. The specific heat of exhaust gases with temperature $t_{FG_{B_o}} = 204°\mathrm{C}$ (477.15 K) and the percentage by volume of products, as well as the specific heat of each gas, should be obtained [10]:

$$c_{p\mathrm{FG}} = c_{p\mathrm{CO_2}} \times \%V_{\mathrm{CO_2}} + c_{p\mathrm{H_2O}} \times \%V_{\mathrm{H_2O}} + c_{p\mathrm{N_2}} \times \%V_{\mathrm{N_2}} + c_{p\mathrm{O_2}} \times \%V_{\mathrm{O_2}}$$
$$= 1.3864\ \mathrm{kJ/m^3{}_{FG}\ grad}\,.$$

The energy derived on complete combustion per unit volume of natural gas or per unit of polyisoprene product depends on the total flue gases' volume ($V_{\mathrm{FG_F}}$), specific heat ($c_{p\mathrm{FG}}$), and flue gases' temperature ($t_{\mathrm{FG_{B_O}}}$) $q_{\mathrm{FG_F}} = V_{\mathrm{FG_F}} \times c_{p\mathrm{FG}} \times t_{\mathrm{FG_{B_O}}} = 3672.2\ \mathrm{kJ_{FG}/m^3{}_F}$ and $q_{\mathrm{FG_P}} = q_{\mathrm{FG_F}} \times v_{\mathrm{F_P}} = 508.23\ \mathrm{kJ_{FG}/kg_P}$ [7]. The volume of exhaust flue gases per unit of the product rejected to the surrounding in the basic process is $v_{\mathrm{FG_P}} = V_{\mathrm{FG_F}} \times v_{\mathrm{F_P}} = 1.7969\ \mathrm{m^3{}_{FG}/kg_P}$. The fuel consumption is $V_{F_h} = D_{P_h} \times v_F = 415.2\ \mathrm{m^3{}_F/h}$. The volume of the exhaust flue gases per hour rejected to the surrounding in basic process is $V_{\mathrm{FG_h}} = V_{\mathrm{FG_F}} \times V_{Fh} = 5390.95\ \mathrm{m^3{}_{FG}/h}$.

3 Solar Energy Application

This case study applies solar-based thermal technology in polyisoprene production for the combined system on location with a collector field of $12.936 \times 10^3\ \mathrm{m^2}$ as seen in Fig. 1. The useful daily solar radiation (Q_u) can be calculated from the monthly average daily radiation (Q_r) and solar collector efficiency (η_c), which differ during the 1-year span from 65% during summertime and 50% in transition to 25% in the wintertime $Q_u = Q_r \times \eta_c$ (kJ/m^2 daily). The required solar collector area is expressed as: $A = F \times Q_{\mathrm{BAC_d}}/Q_u = 1.25 \times 131.73 \times 10^6/Q_u\ (\mathrm{m^2})$, where $F = 1.25$ is the security factor for the large collector installation and $Q_{\mathrm{BAC_d}} = 131.73 \times 10^6$ kJ/day is the daily heat input to the boiler in the plant with air preheater and condensate heat recovery.

The mean flat plate collector area is calculated on the basis of the average values for the area during the summer period, as seen in Table 1 (June, July, August):

$$\overline{A} = A_{6-8}/3 = 37.71 \times 10^3/3 = 12.57 \times 10^3 \rightarrow \left(\mathrm{m^2}\right).$$

In the process with feed water and combustion air preheating with condensate heat recovery, the outlet flue gases temperature was calculated as $t_{\mathrm{FG_{ECO,AP_O}}} = 45.11\,°\mathrm{C}$ (318.26 K).

The monthly useful solar energy with mean collector area can be received using the following equation:

$$Q_m = Q_u \times \overline{A} \times d_m \rightarrow (\mathrm{kJ/monthly})\,.$$

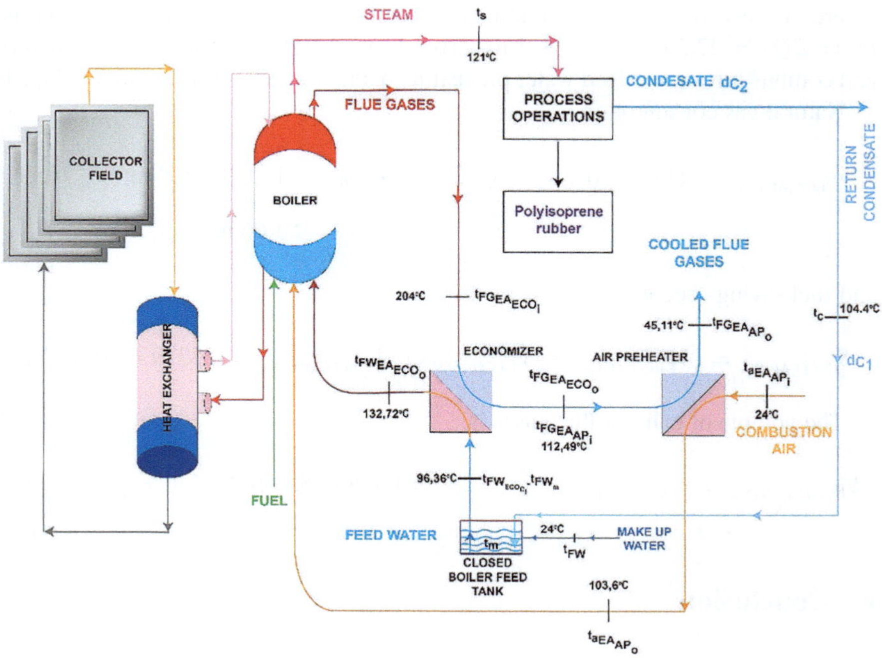

Fig. 1 The solarized process of typical polymer production with condensate heat recovery and feed water/air preheating using flue gases

Table 1 The installed field of the flat plate collectors is located at the location (φ = 45°49′), and the total useful radiation on the horizontal surface is calculated

Month	Average daily radiation, Q_r (10^3 kJ/m^2)	Collector efficiency, η_c (%)	Useful daily radiation, Q_u (10^3 kJ/m^2)	Collector area, A (10^3 m^2)	Days per month, d_m (days)	Monthly useful energy, Q_m (10^9 kJ/month)
February	6.5	25	1.63	101.02	28	0.574
March	9.7	25	2.43	67.76	31	0.947
April	14.8	50	7.4	22.25	30	2.790
May	19.3	50	9.65	17.06	31	3.760
June	20.6	65	13.4	12.29	30	5.053
July	21.3	65	13.85	11.88	31	5.397
August	18.7	65	12.16	13.54	31	4.73
September	14	50	7	23.32	30	2.64
October	8.6	25	2.1	78.41	31	0.818
November	3.6	25	0.9	182.96	30	0.339
$Q_y = \Sigma Q_m$						27.594×10^9

where d_m are days in the month. The total useful yearly solar radiation is: $Q_y = \Sigma Q_m = 27.594 \times 10^9 \rightarrow$ (kJ/yearly). Process with condensate heat recovery and combustion air and feed water preheating using flue gases is presented in Fig. 1.

Natural gas consumption is:

$$V_{F_{ECO,AP,C,S}} = \left(Q_{BECO,AP,C,S_y} - Q_Y\right)/H_L \times \eta_B = 214.63 \times 10^3\ \mathrm{m^3}_{F_{CAS}}/\mathrm{year}$$
$$= 53.66\ \mathrm{m^3}_{F_{CAS}}/\mathrm{h}$$

and fuel savings becomes:

$$S_{F_{ECO,AP,C,S}} = \left(V_{F_{ECO,AP,C,S_h}} - V_{F_{ECO,AP,C,S}}\right)/V_{F_{ECO,AP,C,S_h}} = 0.8379 = 83.792\%$$

The volume of exhaust flue gases is:

$$V_{FG_{ECO,AP,C,S}} = V_{FG} \times V_{F_{ECO,AP,C,S}} = 12.984 \times 53.66 = 696.72\ \mathrm{m^3}_{FG_{ECO,AP,C,S}}/\mathrm{h}$$

4 Conclusion

Through the last few decades, concerns have been raised regarding greenhouse emission during the fossil fuel combustion process. No matter how the industry evolves through digitalization and inventions, the majority of polymer industrial production still rely on classical fossil fuels. Many scientific attempts have been made in order to facilitate renewable energy options to speed up the clean energy transition and reduce production cost. In the middle of the global COVID-19 pandemics, wars, and supply chain interruptions, more concerns have been raised regarding fluctuating fossil fuel's market price. There is a wide range of renewable energy sources, which are suitable for optimizing energy processes in the textile industry sector. The greatest energy savings were shown when combining all of the energy sources together, i.e., the solarized process with flue gases' heat recovery using an air preheater combined with economizer for feed water preheating together with condensate heat recovery. The environmental analysis showed the greatest reduction in natural gas consumption by up to 83.79%. The flue gases' emission is reduced from 5390.95 $\mathrm{m^3}_{FG}$/h to 696.72 $\mathrm{m^3}_{FG}$/h while the gases' exhaust temperature is diminished from 477.15 K (204 ° C) to 318.26 K (45.11 ° C).

References

1. Johnson, I., Choate, W. T., & Davidson, A. (2008). *Waste heat recovery. Technology and opportunities in U.S. industry, technical report*. BSC Inc.
2. Goswami, D. Y., & Kreith, F. (2016). *Energy efficiency and renewable energy handbook* (2nd ed.). Taylor & Francis Group LLC.

3. Mihelić-Bogdanić, A., & Budin, R. (2009). Impact of heat recovery and resource diversification in industrial process; Chapter 12. In *DAAAM international scientific book*. DAAAM International.
4. Afsar, C., & Akin, S. (2016). Solar generated steam injection in heavy oil reservoirs: A case study. *Renewable Energy, 91*, 83–89.
5. Karagiorgas, M., Botzios, A., & Tsoutsos, T. (2001). Industrial solar thermal applications in Greece: Economic evaluation; quality requirements and case studies. *Renewable and Sustainable Energy Reviews, 5*(2), 157–173.
6. Brown, L. H., Hamel, B. B., & Hedman, B. A. (1996). *Energy analysis of 108 industrial processes*. Fairmont Press.
7. Mihelić-Bogdanić, A., & Špelić, I. (2022). Energy efficiency optimization in polyisoprene footwear production. *Sustainability, Special Issue: Sufficiency, Efficiency and Renewable Energy for Sustainable Energy Scenarios, 14*(10799), 1–27.
8. Budin, R., Mihelić-Bogdanić, A., & Vujasinović, E. (2007). Cogeneration and heat recovery in the industrial process. *Chemistry & Industry, 56*, 551–555.
9. Turner, W. C. (Ed.). (2012). *Energy management handbook* (8th ed.). Fairmont Press.
10. Bošnjaković, F. (2012). *The science of heat, volume 1*. Grafis.

Part II
Clean Energy Technology and Emission Reduction

Hydrogen Fuel for a Sustainable Aviation

Gaydaa AlZohbi

1 Introduction

The energy and the automotive sectors have already taken steps toward decarbonization. However, the aviation sector is still in the early stage of decarbonization with a rise of GHG emissions. The aviation fuels require highest quality compared to fuel used in heating, and land and sea transportation. Fluidity, cleanliness, volatility, lubricity, non-corrosivity, and stability are the main properties of aviation fuels. In addition, it might not be influenced by high temperature changes. Actually, the fuels consumed by aviation sector are derived from fossil fuels and are mainly kerosene, methane, gasoline, Jet A/Jet A-1, FT synthetic fuel, and biofuels. Figure 1 displays the carbon dioxide emission from aviation sector between 1940 and 2020. It can be seen that the CO_2 emissions increased, and it reached 1.04 billion in 2018. A rise of global aviation fuel usage with CO_2 emissions has been recorded in previous four decennium. The highest growth was recorded in many developing regions, mainly in Asia owing to the fast development of civil aviation. The average rate of CO_2 emissions growth between 1960 and 2013 was 15 Tg CO_2/year. However, the average growth rate between 2013 and 2018 was 44 Tg CO_2/year. The global aviation CO_2 emissions surpassed 1000 million tones for the first time in 2018.

Even though the efficiency of aircraft increases with reducing the fuel burn at 1% per year, the sizes of aircraft fleets are increasing with 4% per year. As a result, the aviation sector is predicted to generate around 24% of CO_2 emissions by 2050, compared to 3% today [1]. The emissions could be reduced to 19% with a hypothetical enhancement of aircraft efficiency to reach 2.5% per year. Thus, the

G. AlZohbi (✉)
Prince Mohammad Bin Fahd University, Khobar, Saudi Arabia
e-mail: galzohbi@pmu.edu.sa

P. Pong (ed.), *Renewable Energy Resources and Conservation*, Green Energy and Technology, https://doi.org/10.1007/978-3-031-59005-4_8

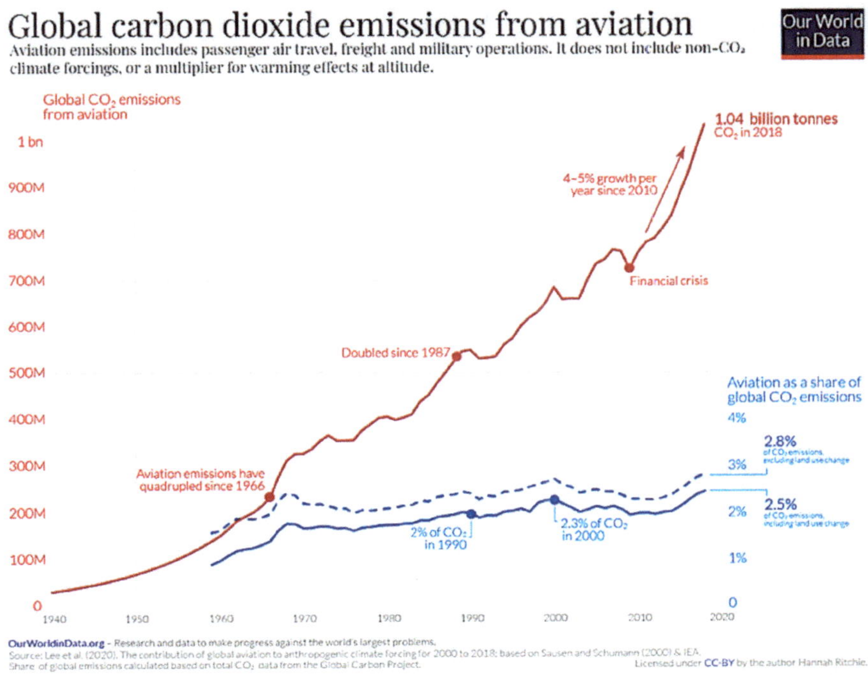

Fig. 1 Carbon dioxide emissions from aviation [2]

aviation industry should start adopting a greener fuel to deal with the predicted high GHG emissions.

Making aviation more sustainable requires a reduction of GHG emissions from flights, cancel out inevitable emissions, and foster a model to lessen air travel. Five options have been identified/considered by aviation industry stakeholders to minimize the GHG emissions from aviation sector by 2050 [3]. These five options are: efficiency gains, hydrogen propulsion technologies, battery, CO_2 offset, and sustainable aviation fuel (SAF). The efficiency gains option is based on improving operation and design to minimize the conventional jet fuel use. Hydrogen option is based on hydrogen combustion or on converting it into electricity through fuel cell. Battery option is established on the use of electric propulsions power using battery charged by a green or zero emission energy resource. CO_2 offsets option is formed on investment in durable out-of-sector CO_2 emissions diminution or elimination. SAF option is fuels generated from sustainable resources such chemical and physical properties as conventional jet fuel. SAF includes: biofuels, generated from biomass or waste; advanced biofuels that are fabricated from biomass and feedstock; and Power-to-liquid fuels (e-Kerosene) that are synthetized by Fischer-Tropsch process, using hydrogen and CO_2 obtained from biomass, industries, or immediately from the air [4].

Sustainable aviation fuels have been proposed by the International Civil Aviation Organization and is considered as a significant measure to minimize GHG releases in aviation sector. SAFs have the benefits of being drop-in-fuels, meaning no need for a change of aircraft and fuel infrastructure. According to [5], SAFs could be incorporated in up to 50% of the fuel ability of the present aircrafts. Thus, SAFs could play a crucial role in decarbonizing the aviation sector. More studies should be conducted to develop an aircraft run with 100% SAF with the ability to eliminate the remaining CO_2 emissions. E-kerosene is based on combining carbon dioxide with hydrogen, generated from water electrolysis. It is adaptable and suitable to be used in traditional fueling framework and combustion engines. E-kerosene might be considered as close to zero GHG emissions if the energy used to run the electrolysis of water is supplied by renewable resources. In addition, e-kerosene generated via Power-to_Liquid technology uses lower quantity of water without using feedstock, enhancing its sustainability aspect [6].

The current chapter aims at reviewing the use of hydrogen as an alternative aviation fuel, focusing on energy consumption, emission-linked cost, and environmental cost. In addition, the different challenges facing the use of hydrogen in aircraft are presented and discussed.

2 Hydrogen in Aviation Sector

Hydrogen is the most abundant element on earth, and it is considered as a clean alternative to methane. Hydrogen fuel could be used directly by injecting it into the natural gas supply network, or by combining it with oxygen in a fuel cell to convert it into electricity. Recently, hydrogen is gaining attention to be used as sustainable fuel for aviation. The main benefits of using of hydrogen in aviation are that it would release only water, and planes can be as rapid as traditional planes, transporting hundreds of travelers along thousands of kilometers. In addition, it contains 2.5 more energy per kg compared to Kerosene [7]. From an environmental perspective, combustion of hydrogen releases less nitrogen oxide, up to 90% less compared to Kerosene, with no formation of particle matter [8]. Liquid hydrogen has high energy density, and it is clean, inexhaustible, and independent of foreign hands [7]. However, the main drawback of using hydrogen is the low volumetric energy density, meaning a requirement for a bigger space to store the hydrogen. The space required for hydrogen is four times more compared to that of Kerosene [9]. The advantages and the disadvantages of using hydrogen in aviation sector in terms of combustion, emission, storage, safety, airplanes, and cost are summarized in Table 1.

Hydrogen can be used in gaseous state, but in order to improve the energy density by volume, hydrogen should be liquified and stored in cryogenic tanks at temperature −253 °C. Liquid hydrogen and gaseous hydrogen are usually transported through road tankers, and pipelines or tube trucks, respectively. The transportation of hydrogen depends on transportation distance and volumes. Truck-

Table 1 Advantages and disadvantages of hydrogen fuel

	Advantages	Disadvantages
Combustion	High mass energy density	Low volume energy density
	High flame speed	
	High combustion temperature	
	Lower flame emissivity	
Emission	Zero emission of CO, CO2, Sulfur, odor	Large emissions of water vapor
	Less emission of NOx	
Storage	Higher thermal stability	Low volume density
	Zero erosive and corrosive contaminants	Low boiling point
Safety	Buoyant vapor	Wide flammability range
	Zero oxygen in fuel system	Very low minimum ignition energy
Airplane	Smaller engine	Bigger tankage
	Less engine waste	Taller fuselage
	Smaller wing area	
Costs	Cheaper aircraft costs	Higher fuel generation capital costs
	Longer lifetime of engine with less maintenance costs	

ing is recommended for small volumes and spitting distance, and pipelines are preferred for larger volumes and longer distance.

Many investigation studies have been already conducted to examine the effectiveness of using hydrogen in aviation sector. The feasibility of using hydrogen in preference of some mission fuel without the need of significant changes of aircraft, and with considering only unutilized luggage space in the lower-deck cargo chambers of airplanes have been studied by Turgut and Rosen [10]. Many parameters could minimize the environmental effects and increase the weight such as container kind, seat capacity, yearly take-off and landing cycles, traveler and luggage load factors, air jet model, and prices of gaseous hydrogen storage, and metal hydride for different sizes. According to [10], despite the rise of cost, there is a substantial decrease of CO_2 emissions, by 25,000–570,000 tones yearly, in many cases and by up to 1.1 million tones yearly for some air jet.

3 Architecture Using Hydrogen Power for Aircraft Thrust

Actually, two designs for hydrogen power for aircraft thrust are considered: hydrogen combustion aircraft and hydrogen fuel cell aircraft. The working principles of the two architectures are displayed in Fig. 2 and explained below. The two architectures are in development stage. The main benefit of the two architectures is the complete removal of CO_2 emissions during flight. Direct hydrogen combustion emits NOx and water vapor, but fuel cells emit only water vapor and are more performant. Besides, hydrogen fuel cell aircraft could offer many features such as

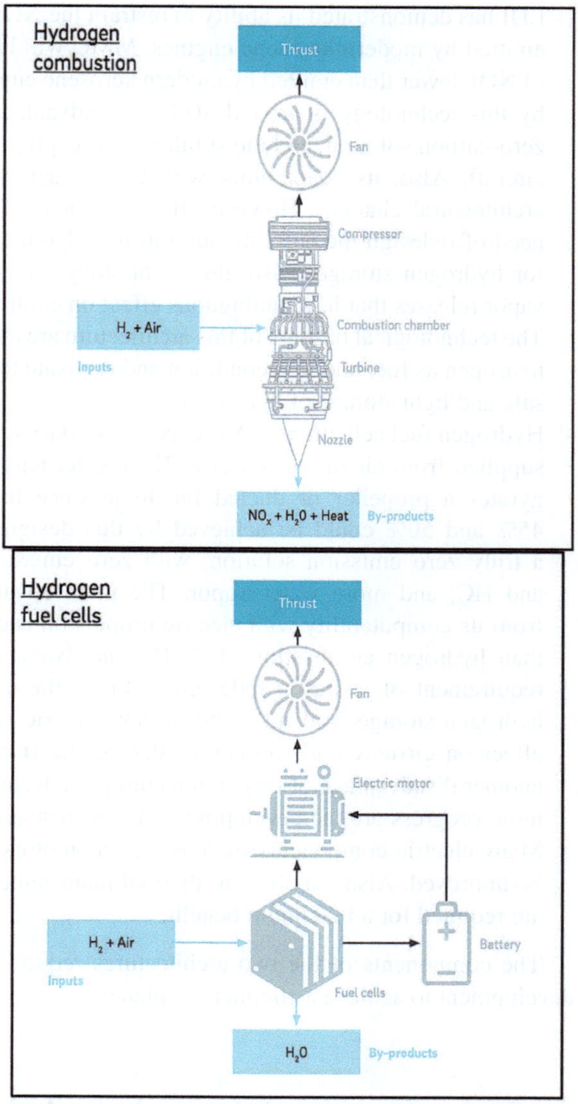

Fig. 2 Schematic illustration of (**a**) hydrogen combustion aircraft and (**b**) hydrogen fuel cells aircraft [11]

electrically propelled aircraft, like electric motors, availing from liability with the quickly electric power train supply chain.

– Hydrogen combustion aircraft: An adapted jet engine is used to produce thrust via hydrogen combustion. In this process, many emissions such as CO, CO_2, and SO_x, produced by conventional jet engines, are removed. However, the emission will be only water vapor and NOx that contribute to GHG emissions. To minimize the emission of NOx, there are two combustor designs being designed and tested: Lean Direct Injection (LDI) and Micro-Mix Combustors (MMC).

LDI has demonstrated its ability to restrain the NOx emissions to the same level emitted by modern kerosene engines. MMC will be able to generate an amount of NOx lower than emitted by modern kerosene engines. The efficiency achieved by this technology is around 40%. The advantages of this architecture are a zero-carbon solution, and the similarity of its propulsion system to conventional aircraft. Also, its adaptability with current aerospace supply chain with less architectural changes. However, the disadvantages of this architecture are the need of redesign the current's aircraft to lodge the supplementary volume need for hydrogen storage. Also, this technology emits NOx and the rise of water vapor releases that have ambiguous effect on contrails/cirrus cloud construction. The technological barriers of this architecture are the need to remodel engines for hydrogen as fuel and to recondition and renovate the aircraft structure to lodge a safe and light storage of hydrogen.

- Hydrogen fuel cell aircraft: A fuel cell is used to transform hydrogen and oxygen supplied from air into electricity. Then, electricity generated runs a motor that gyrates a propeller or ducted fan to produce thrust. An efficiency between 45% and 50% could be achieved by this design. This architecture can offer a truly zero emission solution, with zero emission of CO, CO_2, SO_x, NOx, and HC, and more water vapor. The main advantages of this design derive from its computability with electric propulsion and a 20–40% more performant than hydrogen combustion [12]. The disadvantages of this technology are a requirement of a drastic redesign to lodge the dispersed propulsions system, hydrogen storage, and full suite of new electric subsystems. Also, the unclear effect on cirrus cloud formations due to the rise of water vapor emission is another disadvantage of this architecture. The technological barriers that require more progress are the development of a performant and power dense fuel cells. Many electric components such as electric motors and power electronic should be improved. Also, an efficient thermal management with a new aircraft design are requited for a maximum benefit.

The components of the two architectures require a sufficient and an adequate development to achieve a commercial plane.

4 Environmental and Economic Analysis of Using Hydrogen in Aviation Sector

A comparison between hydrogen-based sustainable aviation fuel compared to kerosene is presented in Table 2. The combustion of hydrogen requires lesser temperature than Kerosene, resulting in lower NOx emissions [13]. The outputs of the hydrogen combustion are thick and thinner ice crystals in comparison to those obtained from kerosene combustion, leading to less grave warming effect. The use of hydrogen, either as combustion or fuel cells, releases 150% more of water vapor than kerosene, meaning less impact on climate since the climate effect on water

Table 2 A comparison between hydrogen-based sustainable aviation fuel compared to Kerosene [15]

% Reduction compared to kerosene	CO_2	NOx	Water vapor	Contrails
Hydrogen fuel cell	−100%	−100%	+150%	−60% to −80%
Hydrogen turbine	−100%	−50% to 80%	+150%	−30% to −50%
E-kerosene generated with hydrogen and CO_2 captures from the air	−100%	−0%	−0%	−10% to −40%

vapor is ten times lower than CO_2 [14]. Water vapor is considered as GHG generated by the combustion of fuel.

An evaluation of the individual and cumulative impacts of the emissions of hydrogen airplanes to kerosene airplanes is performed by Ponater et al. [16] in order to assess the potential of hydrogen airplane in reducing the effect of climate change. Results revealed that the zero CO_2 emission could balance the rise of water vapor exhaust. In addition, CO_2 has a life time of 100 years, which is larger than that of water vapor, that ranges from few days to 1 year [8].

Regarding condensation trails, ice crystals have no place to nucleate because of the lack of solid particles at the discharge of the engine during hydrogen burning process, resulting in a smaller number of water crystals developed at the exhaust. Even so, the crystals that nucleate will have a bigger size owing to the raised quantity of water vapor exhaust. A reduction of the radiative forcing impact of contrails would be predicted. According to Ponater et al. [16], the use of LH_2 airplanes could reduce the radiative forcing by 20–30% by 2050 and by 50–60% by 2100.

Pereira et al. [17] have examined the use of alternative aviation fuels, liquid hydrogen generated by various sources (steam methane reforming, wind, PV, and hydro), and liquid natural gas (LNG) in terms of well-to-wake energy consumption, emissions of CO_2 and local pollutants like, PM, NOx, CO, and HC. Results revealed that LH_2, generated by fossil fuels (SMR) and used in aircraft, consumes 8% less of energy compared to the same aircraft with jet fuel A. However, the use of hydrogen generated by electrolysis that performed by renewable energy is viewed as the best option for high minimization of environmental effects. A reduction range between 51% and 60% of environmental cost and 19% of energy consumption are recorded by using an aircraft fed with LH_2 generated by electrolysis with power supplied by hydro energy. In addition, the amount of fossil fuels consumed by LH_2 from hydro energy is 80% and 84% less than jet fuel A and LNG, successively.

The flight emissions and price of fuel for three types of hydrogen (Gray, Blue, and Green) compared to jet fuel A-1 are presented in Table 3. It can be seen that the green hydrogen (generated by electrolysis operated by renewable energy sources) has zero flight emissions, while it has the highest cost of fuel of 5.96 $/kg. Blue hydrogen (accompanied with CCS) has lower emissions compared to gray hydrogen and Jet Fuel A-1.

Figure 3 displays a comparison between hydrogen, electric battery, and sustainable aviation fuels in term of CO_2 generation, aircraft impact, airport impact, and fuel cost. Battery-electric and hydrogen release less amount of CO_2 compared to

Table 3 Emissions by air jet and fuel cost for different types of hydrogen

	Green H_2	Blue H_2	Gray H_2	A-a jet fuel
Flight emission (metric tons CO_2)	0	57	109	122
Price of fuel ($/kg)	5.96	2.27	2.08	0.56
Passenger emissions (kg CO_2/passenger-mile)	0	0.083	0.157	1.163

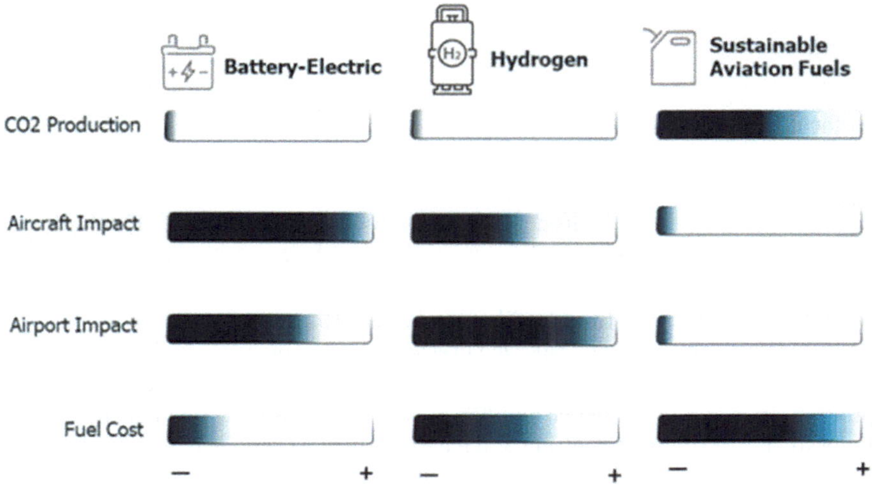

Fig. 3 Comparison of the different propulsion alternative in aviation sector [18]

SAFs. Also, battery-electric has the cheapest fuel cost compared to hydrogen and SAFs. However, battery-electric is viewed as a feasible solution for short distance flight and restricted load/cargo. More investment is required for hydrogen and electric battery regarding airport and aircraft installation. This is due to the need of installing new distribution and storage networks at airports, in fulfillment of new designs of air jet. By contrast, SAEs don't entail significant change given that same systems are used.

5 Challenges Facing the Use of Hydrogen in Aviation

There are many challenges facing the use of hydrogen in aviation sector and require solutions. These challenges are:

– Redesign of engine and aircraft: All components of the aircraft, starting from the propulsion system to the fuel storage, should be redesigned. A fractional redesign of the aircraft is required in the case of hydrogen combustion aircraft, while a complete restructure and remodel are required for hydrogen fuel cells aircraft. The redesign required in the case of hydrogen combustion is based on changing conventional thrust system. In the purpose of lessening the volumetric density relative to jet fuel, changes of fuel delivery and storage, with an extra fuel tank in the fuselage should be performed [19, 20]. These changes could be made through rising the size of fuselage, producing supplemental drag, or through a restructuring of the aircraft, like redesigning the blended wing frames, with a considered closed storage volume. For the hydrogen fuel cell aircraft, a restructure of the thrust is required in order to incorporate a distributed electrical propulsion, including high power/high voltage power electrical systems [21]. Besides the storage requirements, an entire modification of contemporary tube and wings design are needed for the operation of the aircraft.

- Hydrogen storage: As known, the low volumetric energy density is one of the main drawbacks of hydrogen that should be addressed. The space needed to store liquid hydrogen at −253 °C is three times more than needed for kerosene [22]. Thus, efficient storage solutions are required. Liquifying hydrogen is viewed as a promising solution to improve the volumetric energy density. However, the liquid hydrogen requires a cryogenic cooling for a temperature below −253 °C that consume around 45% of the stored energy content [23]. This means an important energy loss between energy stored and energy supplied to thrust, which is defined as tank-to wing efficiency. Thus, the design should be based on retaining high volumetric energy density and high tank-to-wing efficiency. Regarding the cryogenic needed, incorporation of cooling system with an important insulation will be required. To address these requirements, a complicated and weighty tank is required, resulting in minimizing the effective gravimetric energy density of the fuel. Therefore, the redesign should focus on developing light-weighting storage tanks with adequate cryogenic cooling systems.
- Infrastructure: Hydrogen infrastructure should be implemented to enable profiteering the use of hydrogen by aviation sector. The implementation should involve three areas: transportation of fuel to airports, airport refueling infrastructure, and an infrastructure to liquify the hydrogen on the spot. Existing gas networks could be used to transport and deliver fuel. The H21 Leeds city study aims to assess the technical and economic feasibility of transforming the natural gas network to 10% hydrogen gas network in Leeds city in the UK. The output of this study was a roadmap of tasks required for a successful conversion. This study proved the possibility of a conversion of natural gas network into hydrogen gas network with a requirement of an important investment. Liquefaction hydrogen needs a credible and sustainable grid connection to avoid any network disruption cost. In addition, the transportation of hydrogen between the generation location and the airport that could be over a long distance should be considered.
- Cost: Hydrogen is costly compared to the cost of Kerosene. The average production cost of green hydrogen and gray hydrogen are 0.14 $/kWh and 0.05 $/kWh, respectively, without counting the storage cost. Since green hydrogen is required to achieve a sustainable aviation, its price should be dropped to contest the price of kerosene. The cost of hydrogen is anticipated to decrease due to the high demand from transportation sector and supply increase with renewable energy capacity. In addition, the conducting study to develop and improve electrolyser

and hydrogen storage technologies could, through enhancing efficiency and thus minimizing the energy consumption to perform these processes, contribute to further decreasing of the cost of green hydrogen. Besides the decrease in the price of green hydrogen, the rise of carbon cost through imposing emission sanctions on aviation, results in increase in the cost of burning jet fuel. Thus, a monitoring of hydrogen and kerosene price should be conducted by the aviation industry owing to the importance of reversal in the cost differential between kerosene and hydrogen in boosting the investment in hydrogen.

- Sustainable hydrogen generation: An important growth of green hydrogen generation or blue hydrogen will be required to generate efficient volumes for the aviation in a maintainable way. Currently, the high percentage of hydrogen, around 97%, are generated from fossil fuels through steam reforming and coal gasification. However, the generation of green hydrogen through electrolysis accounts only 4%. One of the viable ways to increase the generation of green hydrogen will be through energy storage of the surplus of energy generated from renewable resources. These sources, besides the implementation of carbon capture storage driven by taxes, can rise the generation of green hydrogen with a reduction of its price.
- Safety: Hydrogen gas is known as highly combustible and form incendiary mixtures with oxygen and air [24]. Hydrogen is 14 times lighter than the air, which mean that it can knock-over quickly and the vapors ascend and disseminate [25]. The auto-ignition temperature of hydrogen is larger than that of Kerosene (550 °C vs. 220 °C); however, its minimum ignition energy is smaller than other carbon-based fuels, resulting in an ignition with a weaker spark. Hydrogen is known to be invisible and odorless, making it hard to detect seepage. Moreover, the leaking of hydrogen through powers could be possible when the tank is not correctly insulated due to its small molecule, resulting in high risk. Thus, stringent regulations and secure certifications are required prior to use hydrogen as fuel in aviation. Up to now, more than 50 global policies have been registered to boost the development and the execution of hydrogen [26]. The number of existing policies per type are presented in Fig. 4.
- Hydrogen production efficiency and cost: The comprehensive generation efficiency of hydrogen decreases and the cost rises due to many power transformation steps required in hydrogen generation process. For instance, transforming electricity into hydrogen is considered as an excessive step to merely transform it back into electricity in a fuel cell. Conversely, using a battery to run an airplane is viewed as easier and more performant. Thus, making hydrogen combustion cheaper with high efficiency even with several steps are viewed as a challenge for using it in aviation.
- Price of tickets: The sustainability of aviation sector requires a cooperation from customers. The price of ticket will be higher. The willingness of customers to pay more for greener flights and to fly on a new aircraft owing to their awareness of high safety risks are considered as challenges facing the use of hydrogen in aviation. The decrease of the generation cost of green hydrogen with the imposition of taxes would limit the rise of air ticket.

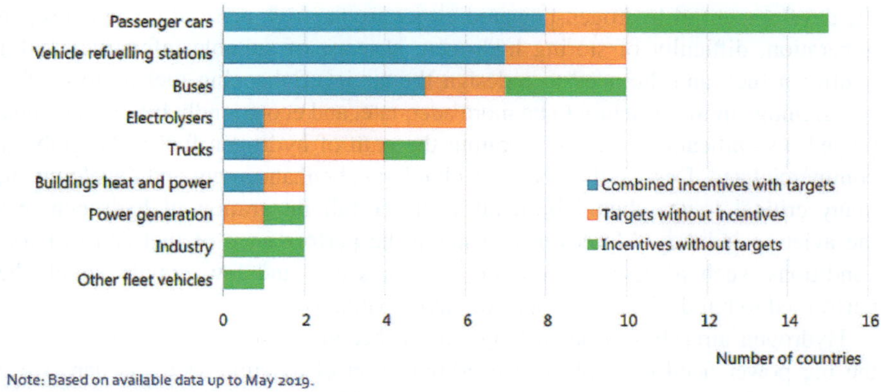

Fig. 4 Number of policies per type [26]

Facing all these challenges requires building a cooperative system between energy suppliers, airline companies, airports, and original equipment manufacturers. An establishment of a worldwide and collaborative system with all the stakeholder is viewed as a leading player to accomplish a sustainable and competitive use of hydrogen in aviation sector.

Airports should construct their own hydrogen generation capacity and establish economies of scale for others uses due to the future emulation between industries for obtaining green hydrogen. Thus, a partnership between airports and energy suppliers should be performed. In order to facilitate the transportation and the use of large amount of hydrogen required for aviation, pipelines will be needed to transport the hydrogen to airports and the liquification process should be on the spot. Moreover, the implementation of infrastructure system should be in different airports in order to smooth the refueling of airplanes flying in difference directions. Also, rules, standardization, and regulations are required to be established between countries to control the air transport and the utilization of hydrogen on the land.

6 Conclusion

The emissions caused by aircraft negatively affect the climate change. These emissions are predicted to rise in the future due to the increase of fuel demand and freight load. Thus, reducing the emissions is crucial to decarbonize the aviation sector. Hydrogen has been considered as a clean and auspicious alternative of traditional aviation fuel. The use of hydrogen in jet engine releases only water vapor, and it removes CO_2 emissions, involving nitrogen oxides, Sulfur, and nitrogen oxides. However, introduction, reforming, and reconditioning of airplanes design to be suitable for hydrogen fuel, besides the related requirement to implement fuel distribution infrastructure are the main snags. The main challenges that slowed down the development of hydrogen-powered air jet are the high price of green hydrogen generation, difficulty of storing hydrogen, absence of suitable infrastructure for hydrogen fuel, and the need to redesign the air jet. Enhancing fuel performance, redesigning engine structure to be more adequate, and eco-friendly by-product have played a significant role in accelerating the path of hydrogen fuel to be globally commercialized. Despite the notable technological enhancement and development, many critical issues should be dealt with for full integration of hydrogen into the aviation industry. Moreover, enhancing the performance of fuel cell in harsh conditions, such as low temperature, low pressure, and low gravity should be performed to build efficient devices for aero-spatial uses.

Hydrogen aircraft still has a long way to become a sustainable reality due to the big power need of airplane. Decarbonization of aviation sector is hard since it requires important development prices and 10 years on average to design a new airplane. The option of carbon capture is a far-off and disproven option. An improvement of fuel efficiency to be adapted for the new design of airplane has been already achieved. Aviation sector is required to consider all existing options to minimize the emissions to be able to halve net CO_2 emissions by 2050 compared to 2005. In addition, a strong collaboration among all players throughout the value chain and across geographic locations is required. In the meantime, alternatives (such as the hybridization of engines) can be explored to reduce carbon emissions.

References

1. Thomson, R. (2020). *Hydrogen: A future fuel for aviation* [Cited 2023]. Available from https://www.rolandberger.com/en/Insights/Publications/Hydrogen-A-future-fuel-for-aviation.html
2. Ritchie, H. (2020). *Climate change and flying: What share of global CO_2emissions come from aviation?* [Cited 2023]. Available from https://ourworldindata.org/co2-emissions-from-aviation
3. Terrenoire, E., Hauglustaine, D., Gasser, T., & Penanhoat, O. (2019). The contribution of carbon dioxide emissions from the aviation sector to future climate change. *Environmental Research Letters, 14*(8), 084019.
4. Bouchy, C., Hastoy, G., Guillon, E., & Martens, J. (2009). Fischer-Tropsch waxes upgrading via hydrocracking and selective hydroisomerization. *Oil & Gas Science and Technology-Revue de l'IFP, 64*(1), 91–112.
5. Enright, C. (2011). Aviation fuel standard takes flight. *ASTM Standardization News, 39*(5), 20–23.
6. Bauen, A., Bitossi, N., German, L., Harris, A., & Leow, K. (2020). Sustainable Aviation Fuels: Status, challenges and prospects of drop-in liquid fuels, hydrogen and electrification in aviation. *Johnson Matthey Technology Review, 64*(3), 263–278.
7. Najjar, Y. S. (2013). Hydrogen safety: The road toward green technology. *International Journal of Hydrogen Energy, 38*(25), 10716–10728.
8. Agarwal, P., Sun, X., Gauthier, P. Q., & Sethi, V. (2019). Injector design space exploration for an ultra-low NOx hydrogen micromix combustion system. In *Turbo expo: Power for land, sea, and air*. American Society of Mechanical Engineers.
9. Verstraete, D. (2009). *The potential of liquid hydrogen for long range aircraft propulsion*. PhD thesis, Cranfield University.

10. Turgut, E. T., & Rosen, M. A. (2010). Partial substitution of hydrogen for conventional fuel in an aircraft by utilizing unused cargo compartment space. *International Journal of Hydrogen Energy, 35*(3), 1463–1473.
11. Khandelwal, B., Karakurt, A., Sekaran, P. R., Sethi, V., & Singh, R. (2013). Hydrogen powered aircraft: The future of air transport. *Progress in Aerospace Sciences, 60*, 45–59.
12. Staffell, I., Scamman, D., Abad, A. V., Balcombe, P., Dodds, P. E., Ekins, P., Shah, N., & Ward, K. R. (2019). The role of hydrogen and fuel cells in the global energy system. *Energy & Environmental Science, 12*(2), 463–491.
13. Sethi, V., Sun, X., Nalianda, D., Rolt, A., Holborn, P., Wijesinghe, C., Xisto, C., Jonsson, I., Grönstedt, T., & Ingram, J. (2022). Enabling cryogenic hydrogen-based CO_2-free air transport: Meeting the demands of zero carbon aviation. *IEEE Electrification Magazine, 10*(2), 69–81.
14. Nøland, J. K., Hartmann, C., & Mellerud, R. (2021). Next-generation cryo-electric hydrogen-powered aviation: A disruptive superconducting propulsion system cooled by onboard cryogenic fuels. *IEEE Industrial Electronics Magazine, 16*, 6–15.
15. SIAPARTNERS. (2021). *The advantages of using hydrogen for sustainable aviation*. Available from https://www.sia-partners.com/en/insights/publications/advantages-using-hydrogen-sustainable-aviation
16. Ponater, M., Pechtl, S., Sausen, R., Schumann, U., & Hüttig, G. (2006). Potential of the cryoplane technology to reduce aircraft climate impact: A state-of-the-art assessment. *Atmospheric Environment, 40*(36), 6928–6944.
17. Pereira, S. R., Fontes, T., & Coelho, M. C. (2014). Can hydrogen or natural gas be alternatives for aviation?—A life cycle assessment. *International Journal of Hydrogen Energy, 39*(25), 13266–13275.
18. Dahal, K., Brynolf, S., Xisto, C., Hansson, J., Grahn, M., Grönstedt, T., & Lehtveer, M. (2021). Techno-economic review of alternative fuels and propulsion systems for the aviation sector. *Renewable and Sustainable Energy Reviews, 151*, 111564.
19. Boretti, A. (2021). Perspectives of hydrogen aviation. *Advances in Aircraft and Spacecraft Science, 8*(3), 199–211.
20. Yılmaz, İ., İlbaş, M., Taştan, M., & Tarhan, C. (2012). Investigation of hydrogen usage in aviation industry. *Energy Conversion and Management, 63*, 63–69.
21. Bradley, T. H., Moffitt, B. A., Mavris, D. N., & Parekh, D. E. (2007). Development and experimental characterization of a fuel cell powered aircraft. *Journal of Power Sources, 171*(2), 793–801.
22. Burkhardt, H., Sippel, M., Herbertz, A., & Klevanski, J. (2004). Kerosene vs. methane: A propellant tradeoff for reusable liquid booster stages. *Journal of Spacecraft and Rockets, 41*(5), 762–769.
23. Aceves, S. M., Espinosa-Loza, F., Ledesma-Orozco, E., Ross, T. O., Weisberg, A. H., Brunner, T. C., & Kircher, O. (2010). High-density automotive hydrogen storage with cryogenic capable pressure vessels. *International Journal of Hydrogen Energy, 35*(3), 1219–1226.
24. Gentilhomme, O., Weinberger, B., & Joubert, L. (2020). Etude de sécurité d'une station de distribution d'hydrogène gazeux. In *Congrès Lambda Mu 22 «Les risques au cœur des transitions»(e-congrès)-22e Congrès de Maîtrise des Risques et de Sûreté de Fonctionnement, Institut pour la Maîtrise des Risques*.
25. Winter, C.-J. (2009). Hydrogen energy—Abundant, efficient, clean: A debate over the energy-system-of-change. *International Journal of Hydrogen Energy, 34*(14), S1–S52.
26. Birol, F. (2019). *The future of hydrogen: Seizing today's opportunities* (Vol. 20). IEA Report prepared for the G.

Experimental Evaluation of a Prototype for the Micro Production of Green Hydrogen

Juan José Milón Guzmán, Mario Enrique Díaz Coa, Damaris Lizbeth Reátegui Herrera, and Rodolfo Caceres Ochoa

1 Introduction

Due to population growth and the expansion of industry and increased, excessive use of fossil fuels such as oil, which may run out in a few years, there are possible environmental consequences such as a greater emission of greenhouse gases and global warming [1–4]. To meet this challenge, the use of green hydrogen represents an interesting alternative fuel since renewable energies are used for its production. The production of green hydrogen is done through the process of electrolysis and it is well-known that there are different hydrogen generation technologies such as by electrodes and/or a dry and wet cell system. But when these technologies work with pure water in the electrolysis process, their hydrogen production efficiency is low due to low electrical conductivity. Therefore, to improve its hydrogen production efficiency, a percentage of catalysts are added, but even here a large amount of energy is lost and it is here that the use of naturally generated energy sources is required in order for costs to decrease. Currently, one of the most important challenges is to reduce the carbonization of the energy industry through the integration of renewable energies into the energy infrastructure, which is why in the year 2022, in research by Domenico Mazzeo et al. [5] they carried out an evaluation of different renewable systems for the production of green hydrogen with the aim of proposing a method to estimate the production of green hydrogen from an energy point of view, whether for wind, solar, or hybrid systems. In 2022, Jaewon Lee et al. [6] conducted research to address the threat of climate change caused by global warming due to CO2 emissions from the use of fossil fuels to generate energy, seeking new technological sources that provide cleaner and more sustainable energy. One of the potentially available sources is hydrogen, which is

J. J. Milón Guzmán (✉) · M. E. Díaz Coa · D. L. Reátegui Herrera · R. Caceres Ochoa
Universidad Tecnológica del Perú, Lima, Peru
e-mail: jmilon@utp.edu.pe

P. Pong (ed.), *Renewable Energy Resources and Conservation*, Green Energy and Technology, https://doi.org/10.1007/978-3-031-59005-4_9

considered to be a clean and sustainable source of energy. However, currently 79% of the total energy is generated through reforming natural gas and coal gasification and 21% is produced as a byproduct of refineries, but these processes also generate considerable CO2 emissions. It is for this reason that in this research they sought to establish a low carbon or carbon-free hydrogen supply chain that is both efficient and low cost. Later, in 2023, Antonio Henriques et al. [7] managed to generate green hydrogen through biomass gasification. At the same time, studies were carried out on three gasification processes: with supercritical water, conventional and plasma, from which production results were obtained with a yield of 0.844 Nm3/kg biomass, 0.828 Nm3/kg biomass, and 0.758 Nm3/kg biomass, respectively. In 2022, in Italy, where there is a large percentage of PV plants and wind plants, an exhaustive analysis of the production of hydrogen through the electrolysis of water was carried out. Angelica Liponi et al. [8] completed a study in which the potential of the production of electrolytic hydrogen throughout Italy concluded with future predictions of a fivefold increase in the capacities of photovoltaic and wind plants by 2029 as compared to 2019.

2 Experimental Approach

The experimental model is made up of: solar energy module (photovoltaic panels, charge controller, and battery bank), hydrogen generation module (hydrogen generator and mass flow meter), and data acquisition module (Fig. 1).

The hydrogen generator module works on the principle of electrolysis; it consists of stainless steel plates separated by insulating polymers immersed in an electrolyte (potassium hydroxide with distilled water). For the supply of electrical energy, the module has two connections (positive and negative). Each solar panel is 370 W, 48 V and has a maximum power current of 9.4 A. The 48 V MPPT charge controller supplies 1920 W. The HIOKI brand current meter model CM 7290 was used to measure the current consumed by the green hydrogen generator module.

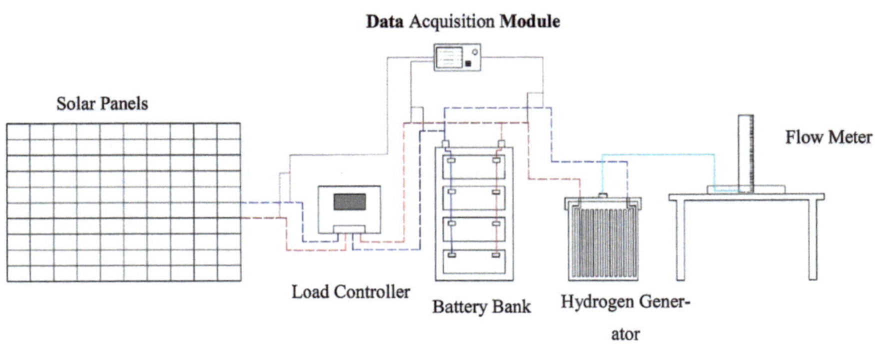

Fig. 1 Design of the prototype for green hydrogen production

The CT7731, which is a 100 A AC/DC current sensor combined with the HIOKI CM 7290 clamp meter display unit, was used to perform the tests. Fluke brand K thermocouple temperature sensors were used, with a measurement range from −40 to 260 °C. This sensor allows monitoring of the temperature variation within the hydrogen generation module in order to find the working range. A measurement plate was used with the objective of adapting the terminals of the measuring instruments with the input of the Keysight model DAQ970A data acquisition system. The data was recorded every 10 s for subsequent analysis. The hydrogen generator consists of 28 conductive plates that go in negative and positive sequence successively. These plates are immersed in a solution of water and potassium hydroxide (KOH).

3 Results and Analysis

3.1 Hydrogen Generation Through the Grid

The graph below shows the production of the hydrogen and oxygen mixture depending on the power (Fig. 2).

Figure 3 shows the production of hydrogen depending on the power.

At this stage, the temperature of the hydrogen generation module was also taken to calculate the appropriate electric current for the production of hydrogen. To do this, the heating speed was calculated for each electric current that was programmed in the charge controller when carrying out the tests. From Fig. 4 it can be interpreted that the heating speed depends on the electric current. This means that the greater the electric current, the greater the rate of heating of the hydrogen generation module, so in order to prolong the operation time it would be better to work with a heat

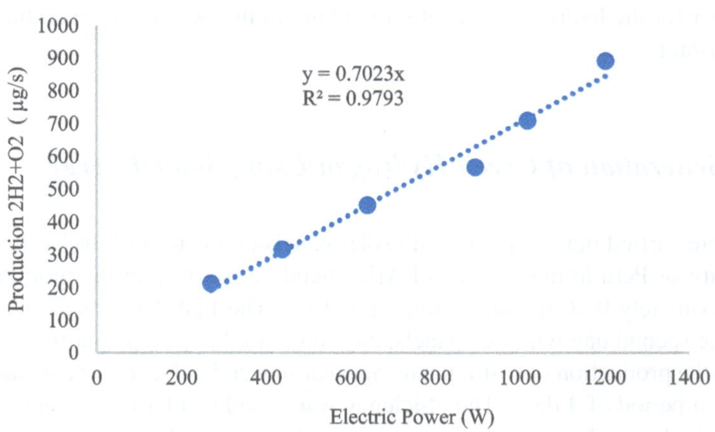

Fig. 2 Production of $2H_2 + O_2$

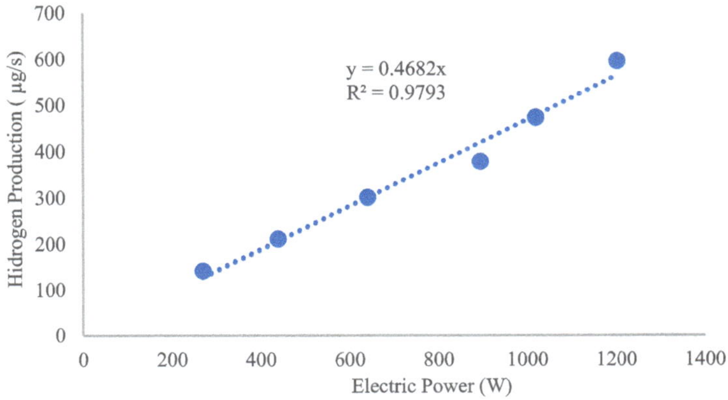

Fig. 3 Hydrogen production in relation to power

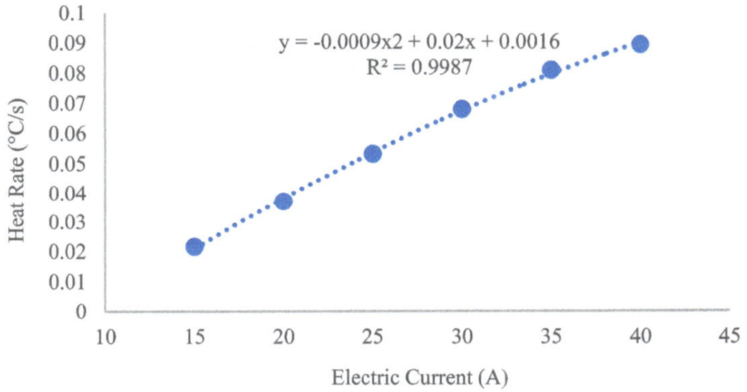

Fig. 4 Behavior of the temperature based on current

exchanger for the hydrogen generation module. In this way the temperature can be kept constant.

3.2 *Generation of Green Hydrogen Using Solar Energy*

Tests were carried out for 4 days with solar panels on the roof of the Technological University of Peru at the Tacna and Arica building, starting in the morning hours at approximately 9:30 h and running until 14 h. The first day was done with one panel, the second day with two panels, and so on until reaching four panels. Figure 5 shows the production of hydrogen using solar power. Each experiment was carried out over a period of 4 days. The efficiency was calculated based on the production of green hydrogen between the power required for its production, where the result was that the green hydrogen generation module has an efficiency of 18.3%.

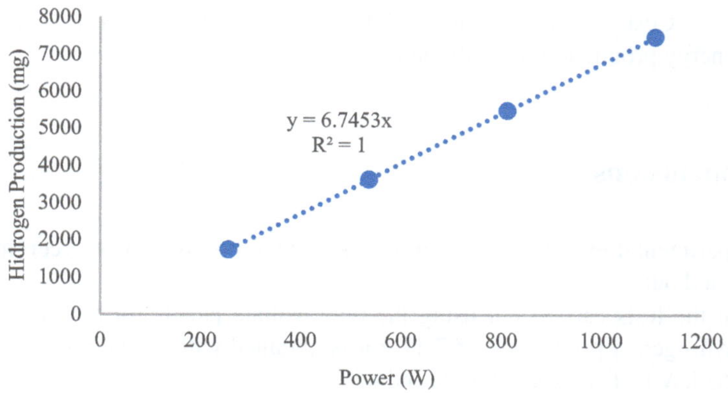

Fig. 5 Production of green hydrogen using solar energy as a source of power

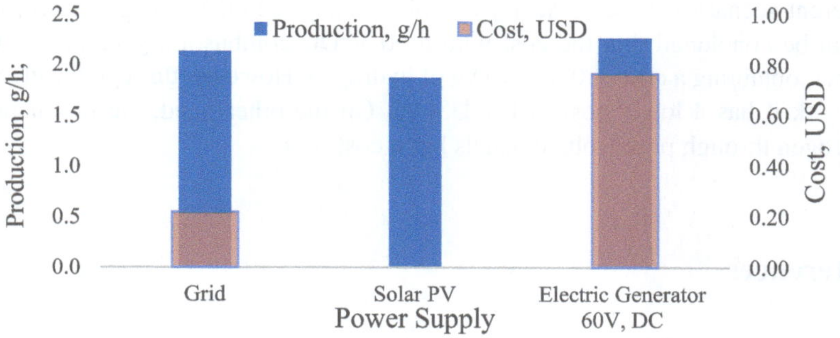

Fig. 6 Comparison of costs in the different production scenarios

3.3 Evaluation of the Cost of Production of Hydrogen by the Prototype

The cost of grid power is 0.22 USD/kW·h and the cost of gasoline is 1.4 USD/L. Figure 6 shows the hydrogen production costs for different scenarios: electrical grid, solar panel, and 60 V DC electric generator. The estimation of production costs was made for 1 h of operation, so in the bar graph it can be seen that the production cost per generator is higher with a cost of 0.76 USD with a production of 2.27 g/h. We can also see the production cost through the Grid, which is 0.22 USD with a production of 2.14 g/h and finally we have the production cost through the solar panels, which is 0.00 USD for 1.87 g/h, making this prototype a good option for isolated areas with good sunlight and where it is difficult to access the electrical grid.

$$C_{\text{Hgrid}} = C_{\text{grid}}.E_{\text{H}} \tag{1}$$

C_{H_grid} = Cost of production, [USD], C_{grid} = Cost of grid [USD/kW·h], E_H = Energy production of Hydrogen [kW·h]

4 Conclusions

The experimental evaluation of a prototype for the production of green hydrogen was carried out.

From the tests carried out using the photovoltaic panels for the generation of green hydrogen, a production of 7.47 g was obtained with an energy consumption of 159.62 kW h. The experiment lasted 4 h.

The efficiency of the prototype for hydrogen production is 18.27%.

The costs of hydrogen production were determined in the prototype in three different scenarios: by electrical grid, solar panel, and 60 V DC generator, where it can be concluded that the cost with a 60 V DC combustion generator is much higher, obtaining a cost of 0.33 USD/g of hydrogen. However, through the electrical network it has a lower cost of 0.1 USD/g. On the other hand, the production of hydrogen through photovoltaic panels has a cost of 0.

References

1. Ganeshkumar, D., & Suresh, M. (2019). A review of performance investigations in hydrogen – Oxygen generator for internal combustion engines. *International Journal of Scientific Research and Engineering Development, 2*(4), 1–6.
2. Barhoumi, E. M., Okonkwo, P. C., & Belg, I. B. (2022). Optimal sizing of photovoltaic systems based green hydrogen refueling stations case study Oman. *International Journal of Hydrogen Energy, Elsevier, 47*, 1–10.
3. Boongaling, A. C., & Bata, K. I. T. (2022). Prospects and challenges for green hydrogen production and utilization in The Philippines. *Revista Internacional de Energía del Hidrógeno, 47*(41), 17859–17870.
4. V. Panchenko, Y. Daus, . A. Kovalev . I. Yud, Prospects for the production of green hydrogen: Review of countries with high potential Hydrogen Energy, vol. 48, 2, p. 4551–4571, 2023.
5. Mazzeo, D., Herdem, M. S., & Mat, N. (2022). Green hydrogen production: Analysis for different single or combined large-scale photovoltaic and wind renewable systems. *Renewable Energy, Elsevier, 200*, 1–19.
6. Lee, J., Ga, S., Lim, D., & Lee, S. (2022). Carbon-free green hydrogen production process with induction heating-based ammonia decomposition reactor. *Chemical Engineering Journal, 457*, 1–45.
7. Henriques Martins, A., Rouboa, A., & Monteiro, E. (2023). On the green hydrogen production through gasification processes: A techno-economic approach. *Journal of Cleaner Production, 383*, 135476.
8. Liponi, A., Pasini, G., Baccioli, A., & Ferrari, L. (2022). Hydrogen from renewables: Is it always green? The Italian scenario. *Energy Conversion and Management, 276*, 1–15.

Stochastic Simulation of Wind Power Profiles from Time Series Analysis Considering Dependencies on Meteorological Variables

Gaia Ceresa, Arianna Trevisiol, Marco Raffaele Rapizza, and Diego Cirio

1 Introduction

The European energy objectives for the power system provide that Renewable Energy Sources (RES) will have to cover 55% of gross electricity consumption in 2030, compared to 35% in 2019; in Italy, this translates into a production of renewable generation equal to 186.8 TWh in 2030, compared to 117.7 TWh in 2019. Therefore, by 2030 it will be necessary to install approximately 40 GW of new renewable capacity mainly from wind and photovoltaic plants. This leads to a less programmable power system than in the past [1]. Therefore, the tools aimed to support various aspects of power system management, from planning to operation, need to evolve to account for the variability and uncertainty of the renewable generation and load. To this aim, probabilistic approaches are increasingly being adopted.

As concerns the long-term planning of the power system, the identification of the need for new projects, and the evaluation of their benefits, in terms of either generation (e.g., to provide investors with suitable market signals to build power plants) or transmission, largely relies on indices obtained by simulating future power system operation.

To account for the variability of system conditions, adequacy (see e.g [2]) and other techno-economic planning indices need to be computed in probabilistic terms. A typical approach is the implementation of the Monte Carlo (MC) method, by which a large number of plausible operating conditions is simulated (MC iterations), each one with input data sampled according to the respective distributions, possibly accounting for dependencies. Different techniques have been used to provide such input data, from conventional historical series analyses to artificial

G. Ceresa (✉) · A. Trevisiol · M. R. Rapizza · D. Cirio
Ricerca sul Sistema Energetico - RSE S.p.A, Milan, Italy
e-mail: gaia.ceresa@rse-web.it

P. Pong (ed.), *Renewable Energy Resources and Conservation*, Green Energy and Technology, https://doi.org/10.1007/978-3-031-59005-4_10

intelligence, passing through complex algorithms for processing the time series. Reference [3] reviews the state-of-the-art in composite system reliability (thus including adequacy), showing various modeling approaches and probabilistic tools for combined generation and transmission system performance evaluation. As far as wind generation is concerned, most of the literature (e.g [4]) addresses the wind power projection from wind speed forecasts made on individual wind farms with a successive spatial aggregation and some reasonable assumptions on wind turbines' technical characteristics. For example, regarding the wind resource in Italy, the study in [5] uses climate model data to elaborate future scenarios of regional power generation until 2100, highlighting overall little variations during the century, with little variations between the regions.

The probabilistic tools that analyze the power system based on MC method may rely on thousands of iterations. To this aim, a procedure called "SPOPSI_wind" (Stochastic wind POwer Profile SImulator) is here presented that generates many plausible hourly year-long time series of wind power, with a spatial resolution at regional scale, starting from historical data and considering the dependencies on temperature and wind speed. To guarantee that each MC iteration has a different input, the series must change from one iteration to another; in addition, they must preserve the typical behavior of the historical wind power series, described by the statistical parameters. The output of SPOPSI_wind can be used directly in the regional tools; it can be aggregated to feed the zonal tools or it can be disaggregated to feed the nodal tools.

The novelty of this procedure is that it generates many wind power profiles, each one with equal probability of realization, and not only the estimated average profile and the confidence region of highest probability of realization. SPOPSI_wind works with aggregated regional data, so that it is less time consuming and occupies less memory compared to the most popular techniques [5, 6], which makes it suitable for tools devoted to the analysis of large networks. Furthermore, SPOPSI_wind uses traditional statistic methods that don't need a huge amount of historical data as required in Machine Learning (ML). Finally, thanks to the dependence on meteorological data, it can generate plausible future series that consider climate changes, if integrated with climate models, though this is being implemented [7]. The chapter is structured as follows: Sect. 2 describes the algorithm; Sect. 3 presents the results. Section 4 concludes.

2 Data and Method

Historical wind power data are from the renewables.ninja data set [8]. They consist of hourly series aggregated at regional level over the whole of Europe, expressed in per unit to be independent of the installed capacity (the evolution of capacity has duly been considered in the setup of the data set, considering a representative wind speed-power characteristic for the wind turbines). In this work, SPOPSI_wind has been applied to the Italian regions only. The historical meteorological inputs are derived from MERIDA HRES (herein called MHRES), a reanalysis data set of the Italian territory at 4 km of spatial resolution, developed by RSE [9]. MHRES provides the atmospheric temperature at 2 m and the wind intensity at 10 m from the ground. The overlapping years between the power and the meteorological datasets are selected, namely, from January 1, 1990, to December 31, 2019.

To model the influence of meteorological data on wind power, it is necessary to combine the different spatial resolution of renewables.ninja power series and of the MHRES meteorological series. For this purpose, first the 851 MHRES cells of Italian municipalities that have at least one wind farm installed in their territory in July 2021 are selected [10]. Then, for each region, one temperature and one wind speed series are computed as an aggregation of the municipal meteorological series: the regional series of temperature and wind speed are the weighted means of the series of the municipalities belonging to the same region and the weight is assumed as the installed capacity. After some tests, the General Additive Model[1] (GAM) is selected as the regression that best models the relation between the monthly mean of the power series and the monthly mean of the meteorological variables.

Given these remarks, SPOPSI_wind, developed in Matlab® and R, is organized in two portions as described in the following.

The first part (analysis), which identifies historical data by the stochastic processes, consists of these three steps:

1. For each region, the monthly average power is computed for all the months in the dataset; then, the original hourly series are divided by this average. The resulting quotients are the "residual."
2. Processing of stochastic residuals of power series:
    (a) The stationarity of residuals is checked with the Augmented Dickey-Fuller and KPSS tests. Then, the residuals are transformed to obtain a Gaussian distribution. Both assumptions are necessary for the stage c.
    (b) The Gaussian series of all the regions are clustered into k groups by applying the k-means algorithm. The most correlated series are thus identified and grouped.
    (c) For each cluster, a Vectorial AutoRegressive model[2] (VAR(p)) is identified.
3. Processing of monthly averages of power series:
    (a) The monthly averages of historical regional temperatures and of wind speeds are computed.

[1] GAM are nonlinear regressions that explain the dependent variable Y (power) as the combination of some smoothing functions s(.) applied to each independent variable x (temperature and wind speed) (used the package mgcv of statistical software R) [11].

[2] VAR(p) is a stochastic process used to capture the relationship between multiple time series as they change over time. Each variable has an equation modeling its evolution over time, that includes the variable's past values, the past values of the other variables in the model, and an error term

(b) The GAM coefficients for each region are identified: monthly average power as dependent variable and meteorological monthly averages as independent ones.

The second part (generation) simulates new series, based on an a priori selected year, necessary to provide the meteorological data, by involving the following steps:

4. The meteorological variables for the selected year are spatially aggregated as seen before in this section, and their monthly average is computed.
5. For each region, the meteorological variables selected and aggregated in step 4 are used as input for the GAM model that generates the simulations of the monthly average power.
6. The VAR(p) models produce stochastic hourly series with Gaussian distribution that are back-transformed into the original distributions of the residuals (inverse of step 2a). They are the stochastic component of the new power series.
7. For each region, the components generated by the GAM and by the VAR(p) models are recombined together. The final series are in per unit of installed capacity.

Remark 1 In this algorithm, the GAM models preserve the dependence of wind power on temperature and wind speed, the VAR models preserve the stochastic behavior (distribution, average, standard deviation, auto- and cross-correlations) and the occurrence of extreme events. Overall, the synthetic series exhibit the typical behavior of the historical ones: their stochastic behavior is close to those of historical series, and this guarantees that they are plausible realizations.

Remark 2 The statistical parameters of synthetic series are not identical to the historical ones, but they are reasonably different as they must reflect the variability of wind power over the years. In fact, the statistical parameters of power production change from one year to another even in the historical series. Furthermore, these series preserve the number of extreme events of the past, i.e., hourly intervals of very low or very high power generation. This is important for adequacy studies of the electrical grid.

Remark 3 SPOPSI_wind does not generate forecasts, but plausible realizations of the process for an arbitrary year, whose meteorological data or climatic projections are known.

3 Results and Discussion

To show the performances of SPOPSI_wind, the first part of the algorithm (analysis) has been carried out with a training set of historical power and meteorological variables of the period 1990–2010; the second part (generation) has been applied by computing 1000 synthetic series of wind power consistent with the statistical

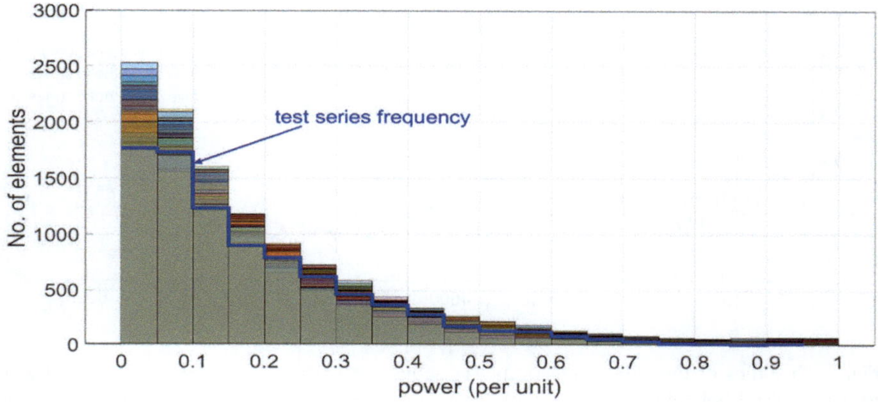

Fig. 1 Histograms of the 1000 simulations overlaid with the histogram of test series (blue line) of power supplied in 2013 in Puglia region

behavior of one of the years belonging to the test set, 2011–2019; the year 2013 was arbitrarily selected.

To validate the proposed algorithm, a comparison between the new simulated series and some test series is shown. As test data, the power delivered in 2013 in all the Italian regions has been considered. For the sake of brevity, the reported results are only about Puglia, the region that has the highest wind power installed in Italy (2.5 GW [10]).

Figure 1 compares the distribution (frequency histograms) of the 1000 new series with that of the test series. The two-sample Kolmogorov-Smirnov test was applied to compare the distribution of each new series against the test one. The result is that the distributions of the new series are almost always recognized as different from the historical one. As anticipated above, this is good because the series must differ between each other, provided that the statistical properties are preserved. In fact, Fig. 1 qualitatively shows that the new distributions are consistent with the historical ones (peak in the first and second bins, decreasing height of other bins).

Figure 2 compares the hourly power profile of the test series (wind power delivered in each hour by all the wind farms in the region, in blue) with the 5th, 50th, and 95th quantiles of the profiles computed from the 1000 new series (red lines). 120 h moving average is applied to all the plotted lines for readability reasons. The test series is included between the 0.05 and 0.95 quantiles, it is well approximated by the median (0.5 quantile) of the new series, and the annual trend is respected.

Figure 3 presents the cumulative sum[3] of all the new power series (some quantiles in red), and the blue line is the cumulative sum of the test series. The test series is

[3] Cumulative sum = sequence of partial sum of a sequence; in a series with N elements, the cumulative sum is $x_1, x_1 + x_2, x_1 + x_2 + x_3, \ldots, \sum_{i=1}^{N} x_i$.

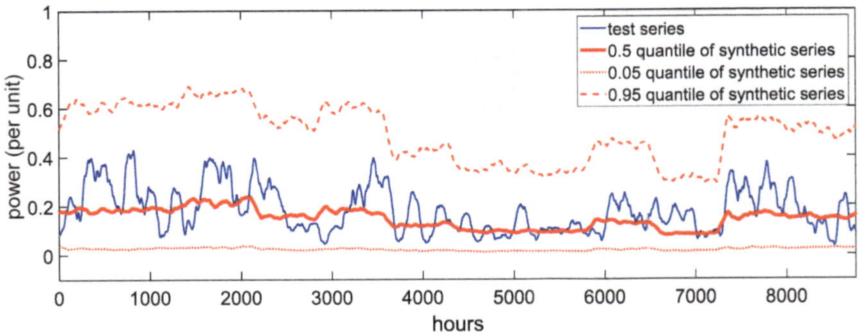

Fig. 2 Quantiles of the 1000 synthetic power series overlapped to the test one (power profile realized in 2013 in Puglia region)

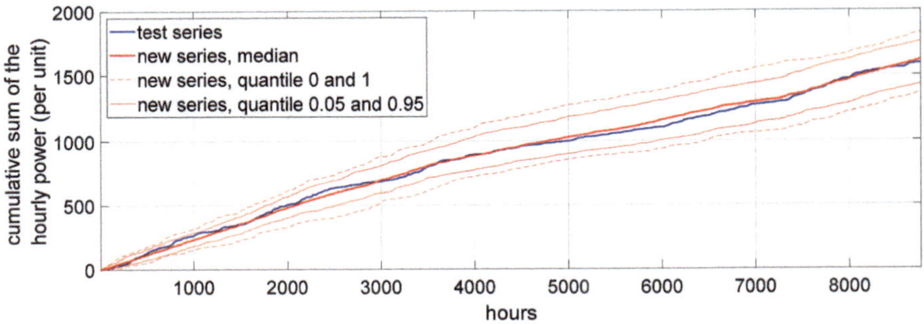

Fig. 3 Cumulative sum of test series (blue, realized in 2013 in Puglia region) and quantiles of the cumulative sum of the synthetic series (red)

close to the median of the new cumulative sums: this means that the variability of the new series is consistent with that of the test one all along the year.

To provide an overview of the results in all the regions, Table 1 reports the average, the standard deviation and the equivalent hours[4] of the historical data and of the synthetic series. In the historical case, for each region, one index for each year is computed (one annual average, one annual standard deviation, and one annual equivalent hour for each year), so there are 30 values for each region for each index: the table shows only the minimum and the maximum. The column called "2013" presents the average, the standard deviation and the equivalent hours of the realized series in the regions in 2013. The columns about synthetic series show the minimum and the maximum statistical indices computed on the 1000 new series generated for each region with the meteorological data realized in 2013. The data are expressed in "per unit." Again, the statistical indexes of the new series prove to be as desired,

[4] Equivalent hours = sum of all the per unit of the annual series.

Table 1 Statistical parameters of historical and new series of wind power

	Average					Standard deviation					Equivalent hours				
	Hist. annual series			Synthetic series		Hist. annual series			Synthetic series		Hist. Annual series			Synthetic series	
Region	Min	Max	2013	Min	Max	Min	Max	2013	Min	Max	Min	Max	2013	Min	Max
Piem	0,07	0,1	0,09	0,07	0,01	0,08	0,12	0,1	0,08	0,11	624	874	757	592	906
VdAo	0,16	0,25	0,22	0,16	0,27	0,22	0,27	0,25	0,19	0,29	1418	2227	1967	1397	2403
Ligu	0,07	0,1	0,08	0,07	0,1	0,08	0,12	0,1	0,07	0,11	600	833	718	576	838
Lomb	0,11	0,16	0,14	0,11	0,16	0,12	0,16	0,14	0,11	0,16	927	1392	1234	994	1443
Tren	0,12	0,17	0,13	0,12	0,16	0,18	0,22	0,18	0,17	0,22	1047	1479	1106	1035	1392
Aadi	0,15	0,2	0,16	0,15	0,19	0,17	0,2	0,17	0,16	0,2	1293	1747	1397	1311	1664
Vene	0,14	0,2	0,15	0,15	0,18	0,15	0,19	0,17	0,15	0,19	1217	1712	1348	1300	1573
FrVG	0,16	0,21	0,18	0,17	0,2	0,15	0,18	0,16	0,14	0,18	1414	1879	1592	1452	1769
EmRo	0,15	0,21	0,17	0,15	0,2	0,16	0,2	0,17	0,15	0,2	1319	1814	1480	1335	1741
Tosc	0,15	0,19	0,16	0,15	0,18	0,15	0,18	0,16	0,15	0,19	1304	1623	1438	1292	1614
Umbr	0,15	0,21	0,18	0,15	0,19	0,16	0,19	0,19	0,15	0,19	1320	1824	1558	1345	1689
Marc	0,16	0,2	0,2	0,17	0,2	0,17	0,2	0,2	0,16	0,2	1402	1753	1715	1451	1749
Lazi	0,1	0,15	0,13	0,1	0,17	0,15	0,2	0,17	0,14	0,22	855	1285	1107	843	1466
Abru	0,12	0,21	0,19	0,15	0,21	0,15	0,22	0,21	0,18	0,23	1014	1836	1645	1346	1821
Moli	0,09	0,11	0,11	0,08	0,12	0,09	0,12	0,12	0,09	0,12	749	959	929	674	1035
Camp	0,1	0,14	0,13	0,09	0,14	0,12	0,17	0,15	0,11	0,16	847	1226	1099	825	1187
Pugl	0,08	0,1	0,09	0,07	0,11	0,08	0,11	0,08	0,08	0,11	667	900	750	624	948
Basi	0,11	0,15	0,12	0,11	0,14	0,11	0,16	0,14	0,11	0,15	975	1312	1084	975	1235
Cala	0,09	0,13	0,1	0,09	0,12	0,11	0,15	0,12	0,1	0,14	797	1124	854	777	1059
Sici	0,11	0,16	0,12	0,11	0,15	0,12	0,17	0,14	0,12	0,17	995	1375	1091	976	1334
Sard	0,11	0,13	0,11	0,09	0,13	0,11	0,15	0,12	0,1	0,15	926	1143	956	797	1177

Historical indexes refer to renewable.ninja annual series of the period 1990–2019 and a focus on year 2013; synthetic series are simulation of wind power production consistent with the historical series of 2013

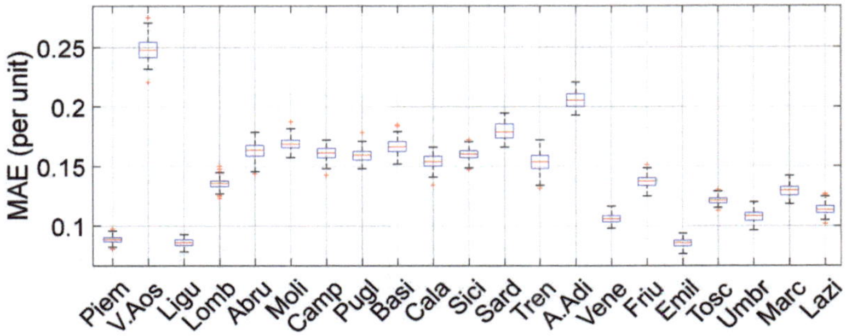

Fig. 4 For each region, representation of the 1000 MAE, computed by the mean absolute differences between the 1000 new series and the test one

i.e., comparable, but not identical to those of the annual series realized in all the available periods.

Finally, Fig. 4 shows the Mean Absolute Error (MAE). In this case it should not be considered as an error, but as a distance between the new series and the test one. For each region, given $s \in [1, 1000]$ new series and $i \in [1, 8760]$ hours of the year, $y_{s,i}$ is the power estimated by the sth new series at the ith hour, $\hat{y}_i$ is the power delivered in the region at the ith hours, so their distance is $\text{MAE}_s = \frac{1}{8760}\sum_{i=1}^{8760}\left|y_{s,i} - \hat{y}_i\right|$. For each region, a boxplot collects all the MAE_s: the box contains the values, in per unit, between the 25th and the 75th percentiles, and the red line is the median. The largest distances (highest MAE) with respect to the test series are in the regions characterized by very few and small wind farms, and a very complex orography. Despite that, the statistical properties are preserved. The other regions have a MAE lower than 0.2.

4 Conclusion

The chapter has presented SPOPSI_wind, a procedure that generates stochastic hourly series of wind power at the regional level starting from historical data of generated power and meteorological variables. As shown in Sect. 3, the new power series preserve the statistical behavior of the historical wind power series, though with different hourly profiles. This is important, because the output of the procedure is aimed to feed tools for probabilistic analysis of electric power systems implementing MC simulation to compute adequacy or other techno-economic indices for long-term system planning. The dependence of the wind power from historical temperature and wind speed is modeled to generate plausible wind power series for future scenarios, with the potential to consider the effect of climate changes. By working with aggregated data, this procedure is faster than the techniques based on detailed wind farm models, thus it can deal with large power systems. In addition, SPOPSI_wind uses traditional statistic methods that need less historical data and computational time than ML approaches. It is versatile because the time series, expressed in per unit, are independent of the installed capacity in a given year. The procedure has been applied to the Italian case featuring only onshore wind generation; however the approach is general and can be used for offshore wind farms as well. As a further development, a post-processing will be addressed to consider wind turbine efficiency increase over time, to modify the generated wind power profiles accordingly. Furthermore, the procedure will be extended to consider other renewable resources and load demand.

Acknowledgments This work has been financed by the Research Fund for the Italian Electrical System under the Three-Year Research Plan 2022-2024 (DM MITE n. 337, 15.09.2022), in compliance with the Decree of April 16th, 2018.

References

1. Terna. (2021). *The national electricity transmission grid development plan*. https://www.terna.it/en/media/press-releases/detail/2021-development-plan-national-electricity-grid-presented. Last accessed 2023/08/11.
2. ENTSO-E: European Resource Adequacy Assessment. (2022). https://www.entsoe.eu/outlooks/eraa/2022/. Last accessed 2023/08/11.
3. IEEE Composite System Reliability Task Force. (2022, Aug). *Composite power system reliability, technical report PES-TR99 IEEE.*
4. Ahmed, S. I., Ranganathan, P., Salehafar, H. (2021). *Forecasting of mid- and long-term wind power using machine learning and regression models*. IEEE Kansas Power and Energy Conference (KPEC), Manhattan, KS, USA.
5. Bonanno, R., Viterbo, F., & Riva, M. (2023). Climate change impacts on wind power generation for the Italian peninsula. *Regional Environmental Change, 23*, 15.
6. FlexPlan Project. (2022). *Monte Carlo scenario generation and reduction D1.1.*
7. Trevisiol, A., Ceresa, G., Bonanno, R. (2023). *Climate series processing for a stochastic procedure to generate on-shore wind power future scenarios in Italy*. SISC 2023 Conference, Milan, 2023/11/22-24.
8. Staffell, I., & Pfenninger, S. (2016). Using bias-corrected reanalysis to simulate current and future wind power output. *Energy, 114*, 1224–1239.
9. Bonanno, R., Lacavalla, M., & Sperati, S. (2019). A new high-resolution meteorological reanalysis Italian dataset: MERIDA. *Royal Meteorological Society, 145*(721), 1756–1779.
10. GSE: Atlaimpianti. (2021). https://atla.gse.it/atlaimpianti/project/Atlaimpianti_Internet.html. Last accessed 2023/08/11.
11. Wood, S. N. (2006). *Generalized additive models, an introduction with R.* Chapman and Hall/CRC.

Short-Term Scheduling of Support Vessels in Wind Farm Maintenance

Manru Xue and Paulo Cesar Ribas

1 Introduction

When we discuss energy mixes today, we consider a wide range of sources, including coal, oil, gas, nuclear, hydropower, solar, wind, and biofuels. Currently, oil provides most of the world's energy consumption, followed by coal, gas, and hydroelectric power. However, three-quarters of global greenhouse gases come from burning coal, oil, and natural gas. This has caused the warming of the Earth and endangered the living environment of human beings. To avoid global warming of 1.5 °C, we must stop at least 80% of all energy and non-energy fossil fuel and biofuel emissions as soon as possible [1].

To prevent further deterioration of environmental problems, it is urgent to find a dry-friendly and sustainable energy source to replace fossil fuels. With the continuous exploration by people, several clean and relatively high-value sustainable energy sources have been discovered, including wind energy. Wind energy is a type of energy produced by air currents. In recent years, it has become common to separate between onshore wind power and offshore wind power. However, the effort to explore the offshore wind power industry has intensified due to certain geographical and environmental restrictions on onshore wind power. Although offshore wind power has become very popular, its industrial cost is also high. This includes not only the high manufacturing and installation costs of the turbines but also the associated high maintenance and operating costs [2]. Among them, the route planning and scheduling of the daily maintenance fleet is one of the critical factors affecting the overall operation and maintenance cost.

We defined mathematical models for creating a maintenance plan for offshore windmills. For example, when there is an offshore wind farm, we need to formulate a

M. Xue · P. C. Ribas (✉)
Molde University College, Molde, Norway
e-mail: paulo.c.ribas@himolde.no

P. Pong (ed.), *Renewable Energy Resources and Conservation*, Green Energy and Technology, https://doi.org/10.1007/978-3-031-59005-4_11

plan to optimize daily operation and maintenance to minimize the total maintenance cost. So, assuming that a large service operating vessel (SOV) can carry two smaller crew transfer vessels (CTV) to complete a 2-week maintenance mission. The SOV stops at different places every day, and the CTVs need to complete the maintenance tasks of the day and return to the SOV according to the plan.

The present chapter proves that it is possible to solve this problem by splitting it into three, using a set of models. Additionally, the chapter demonstrates that the time to run this set of models is acceptable. Thus, our main contribution is to present an efficient and straightforward approach to solve this new problem on O&M Offshore Wind Energy.

2 Literature Review

A large body of literature considers weather uncertainties on offshore wind logistics problems. Stålhane et al. [3] consider it on route optimization of crew transfer vessels that depart from an onshore base. Stålhane et al. [4], Gundegjerde et al. [5], Stålhane et al. [6], and Stålhane et al. [7] consider it in the fleet sizing of O&M vessels.

Irawan et al. [8] considers the expected turbine failure to propose simulation-based optimization to solve the route problem for the crew transfer vessels departing from a base.

Similar to the traveling salesman issue used for offshore wind farm route planning, Stock-Williams and Swamy [9] provides a meta-heuristic optimization method to identify the strengths and weaknesses of any maintenance program and estimate the investment in implementation.

Dawid et al. [10] studied an O&M tool for short-term decision-making that saves time and costs while extending the effective remedy window. Lazakis and Khan [2] create a novel optimization heuristic framework for daily or short-term operations based on route planning and scheduling to reduce costs under various operational constraints. Li et al. [11] developed a decision support system (DSS) to reduce OWF maintenance costs. The DSS is designed to be used by a broad range of stakeholders in the OWF industry to guide maintenance strategies, eventually reducing the overall cost of OWF life cycle maintenance.

Irawan et al. [8] suggested an optimization approach for solving stochastic issues under unknown situations based on SMRP simulation. The model was created to optimize each ship's repair schedule and route for many days, as well as the ship's ability to transfer spare parts. Li et al. [12] provided an optimization methodology to guide long-term maintenance strategies for offshore wind farms under actual settings with significant uncertainty. Compared to earlier research, this technique is established for a more realistic maintenance decision-making environment to quantify the influence of uncertainty on maintenance performance and provide a set of maintenance methods [13]. The newest OWT maintenance research is examined in this study, including strategy selection, schedule optimization, field operations, maintenance, evaluation criteria, recycling, and environmental challenges. This paper analyzes the constraints of OWT operation and maintenance research as well as the lack of industrialization progress while describing and comparing different methodologies.

Manupati et al. [14] offer a novel plasma supply chain model that creates an efficient Mixed Integer Linear Programming (MILP) model by balancing two opposing objective functions. It also serves as the foundation for the model developed in Task 1 of this chapter.

Campuzano et al. [15] provide a simple algorithmic strategy for improving the computation of the Miller-Tucker-Zemlin (MTZ) model of the Asymmetric Traveling Salesman Problem (ATSP) by effectively generating efficient inequalities from fractional solution performance. This strategy is used to generate the model in Task 2 of this chapter.

Different phases of research on the vehicle routing problem with time windows (VRPTW) have been conducted [16, 17]. The models in the literature provide helpful inspiration for developing the Task 3 model in this chapter.

3 Problem and Methodology

3.1 Offshore Wind: Operation, and Maintenance

Offshore wind farms are currently being installed away from the coast in search of locations that meet space and wind conditions to increase energy production [18]. These locations may be suitable for the deployment of wind turbines and have optimal conditions for energy production. However, the logistics of operations and maintenance are becoming increasingly complex. Wind farm operators are under considerable pressure to cut costs to make energy production profitable. Maintenance operations are one of the most expensive components of an offshore wind farm, accounting for up to 25% of the total costs [3].

Due to multiple constraints such as weather, type of failure, and available vessels and technicians, deciding which turbines to maintain, the order to access them, and the vessel's route is challenging. Poor maintenance operations planning on large vessels such as Service Operation Vessels (SOVs) and Crew Transfer Vessels (CTVs) can result in high fuel consumption, increasing overall maintenance costs and, more importantly, contributing to the carbon footprint of offshore wind farms. As a result, planning the day-to-day operation and maintenance of offshore wind farms is a key but complex and challenging problem. To solve this problem, it is necessary to find the best route to maintain the turbines for the Service Operating Vessel (SOV) and Crew Transfer Vessel (CTV) to minimize the total cost.

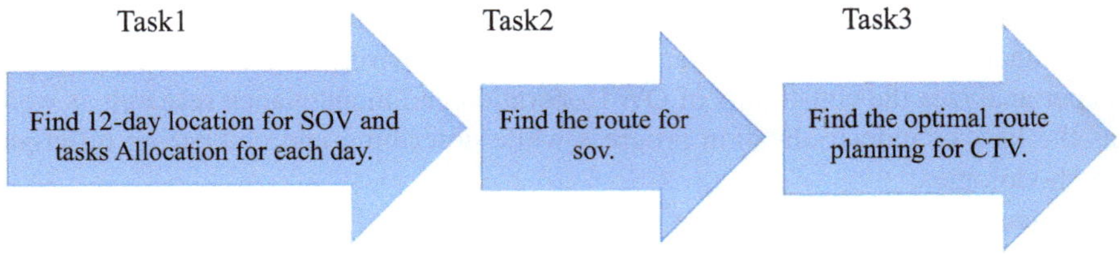

Fig. 1 Solution approach

3.2 Solution Approach

We analyze a maintenance cycle of 12 days (equivalent to one SOV voyage period), using one SOV and two CTVs as maintenance ships, and define a mathematical model to solve with AMPL for finding the best sailing route for daily maintenance ships. Unfortunately, we don't have enough space to discuss and analyze the models here, so we will focus on the results and model performance.

The overall case is divided into three tasks (Fig. 1); for each one, a model was developed:

For Task 1, the mathematical model is created adapting a traditional model of the Location & Allocation Problem [14].

The objective function aims to minimize the distance between the turbines that need maintenance and the turbine where the SOV will stay during the day, so this model has a high impact on the Task 3 model; a good result on Task 1 provides short distances for Task 3. It also considers an approximate maximum capacity per day, ensuring, in this way, the feasibility for Task 3. As a result of this model, we have the location of the SOV for 12 days and the maintenance jobs allocated for each day.

In Fig. 2, we can see a solution example where the orange points are the positions where the SOV will stay each day, and the areas contain the tasks to be performed from this position. Some days, the SOV may perform only one task a day; in this case, there is no area, just the orange point.

Task 2 is used to find the route for the SOV, knowing the solution of Task 1 with the 12 locations. To achieve it, we used a TSP model based on work [15] with sub-cycle prohibiting constraints.

Figure 3 has a solution example where, on the left, it is possible to see the real dimensions of the Dogger Bank case, and on the right, we can see the SOV route during the 12 days between the stay points for each day.

While Task 3 model, based on Task 1 result, makes the route to CTVs for each day. This model was developed based on a Vehicle Routing Problem with Pickups and Deliveries (VRPPD) and linked Time Windows; the objective function is to minimize the distance traveled by the CTVs.

A solution example of this model is shown in Fig. 4, where P is the stay point for the SOV, the yellow line contains the points for the first CTV route, and the black line for the second one.

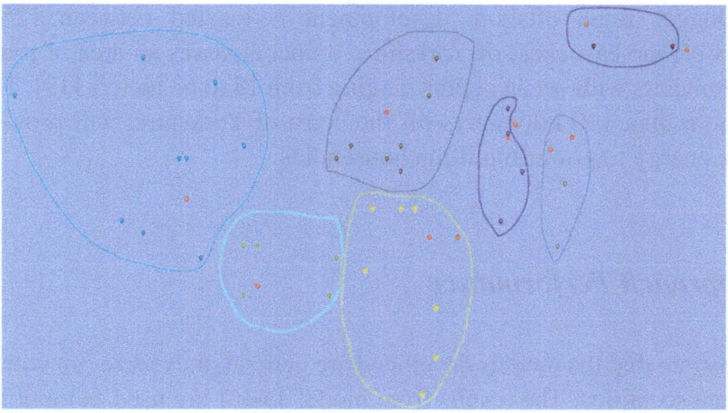

Fig. 2 Example Task 1 solution

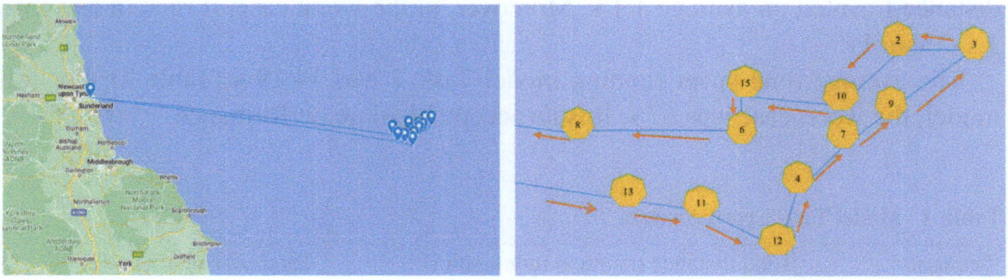

Fig. 3 Example Task 2 solution

Fig. 4 Example Task 3 solution

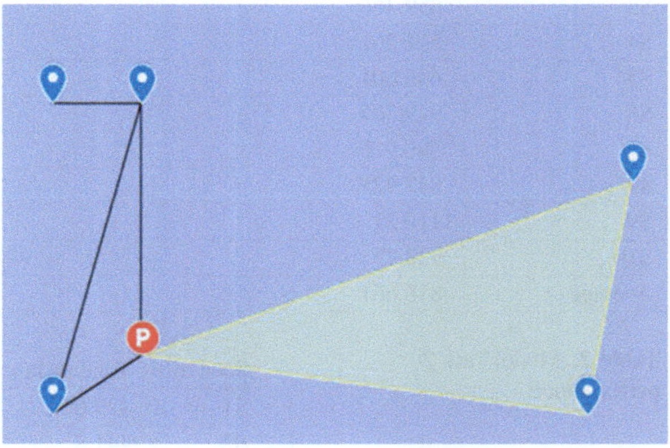

3.3 Case Description

The approach developed in this work was applied to the Dogger Bank wind farm. Some data are real, and some are realistic data based on different real operations and academic papers.

Dogger Bank is a sandbar in the middle of the southern North Sea.

The Dogger Bank offshore development is located between 125 and 290 kilometers off the east coast of Yorkshire. It encompasses an area of roughly 8660 square kilometers with ocean depths ranging from 18 to 63 meters [19]. The project, located more than 130 kilometers off the coast of Yorkshire, will provide enough renewable energy to power 6 million households.

3.4 Approach Performance

To demonstrate that the models developed are general, they were run with data from ten different scenarios. The results from model Task 1 are used as input for the two other models. We used AMPL and CPLEX 20.1.0.0 solver to run all models.

The model Task 1 had an excellent performance, spending to achieve the minimal distance traveled on average 1.1 s. Moreover, in the worst case, it runs in less than 2 s (Table 1).

The average time spent running model Task 2 was 14.19 s (Table 2). For this model, the worst case was 67 s. It is an acceptable time performance.

Table 1 Model Task 1 performance

Scenario	Total distance (nautical miles - nmi)	Total solving time (s)
S1	902.363	0.875
S2	560.819	1.15625
S3	859.466	1.125
S4	574.261	1.46875
S5	646.601	0.484375
S6	949.303	1.3125
S7	796.926	0.6875
S8	847.928	0.921875
S9	1110.21	1.89062
S10	938.73	1.07812
Average	818.661	1.01000

Table 2 Model Task 2 performance

Scenario	Total distance (nmi)	Total solving time (s)
S1	429.619	16.1562
S2	441.8	13.9844
S3	439.268	4.78125
S4	438.259	2.42188
S5	436.8	3.70312
S6	435.64	10.7188
S7	433.752	67.0781
S8	432.417	2.375
S9	429.308	3.78125
S10	434.404	16.9062
Average	435.1267	14.19062

Table 3 Model Task 3 performance

Scenario	Total distance (nmi)	Total solving time (s)	Number of stops
S3	6.80968	0.484375	8 (4 turbines)
S5	19.1683	19.2812	16 (8 turbines)
S7	12.0173	1.45312	10 (5 turbines)
S8	10.0811	0.046875	4 (2 turbines)
S9	8.08198	0.140625	6 (3 turbines)
S10	14.1929	6.21875	14 (7 turbines)
S12	32.0456	5.54688	18 (9 turbines)
Average	14.62812286	4.7366	

Model Task 3, with distance minimization as the objective function and maximal journey time as a limit, spent, on average, 4.74 s to achieve the optimum (Table 3). However, with a high variation, it takes almost 20 seconds for complex scenarios.

4 Conclusions

This work has presented a solution approach for solving the problem of planning the maintenance of offshore windmill parts. It analyzed related research problems through relevant literature research and tested the model developed. After several tests, a satisfactory result was obtained on the test cases based on real-world data.

Through Model 1, we find the minimum value of the distance traveled, including the distance between the turbines assigned to each place. Moreover, found the 12-day stay of SOV and the schedule of turbines to be repaired. Then by Model II, we find the minimum total travel distance of all used edges for all scenarios and SOV loading points and routes for 12 days. Finally, by running Model 3, the distances of the six cases in the CTV pickup and delivery scenarios and the location of the CTV and the SOV based on the route are found.

Splitting the problem was possible to solve it using a speedy performance suit.

4.1 Discussions and Future Works

Despite the proposed approach's good performance, it is still ongoing work. We focused on distance minimization; however, other objective functions should be studied. For example, for Task 1, the balance between the working time allocation in the different days. And for Task 3, the minimization of completion time.

These alternative objective functions can be relevant for the decision maker but probably will have worse performance, and it will demand additional model development.

References

1. Jacobson, M., Delucchi, M., Cameron, M., Coughlin, S., Hay, C., Manogaran, I., Shu, Y., & von Krauland, A. (2019). Impacts of Green New Deal energy plans on grid stability, costs, jobs, health, and climate in 143 countries. *One Earth, 1*(4), 449–463.
2. Lazakis, I., & Khan, S. (2021). An optimization framework for daily route planning and scheduling of maintenance vessel activities in offshore wind farms. *Ocean Engineering, 225*, 108752.
3. Stålhane, M., Hvattum, L. M., & Skaar, V. (2015). Optimization of routing and scheduling of vessels to perform maintenance at offshore wind farms. *Energy Procedia, 80*, 92–99.
4. Stålhane, M., Vefsnmo, H., Halvorsen-Weare, E. E., Hvattum, L. M., & Nonås, L. M. (2016). Vessel fleet optimization for maintenance operations at offshore wind farms under uncertainty. *Energy Procedia, 94*, 357–366.
5. Gundegjerde, C., Halvorsen, I. B., Halvorsen-Weare, E. E., Hvattum, L. M., & Nonås, L. M. (2015). A stochastic fleet size and mix model for maintenance operations at offshore wind farms. *Transportation Research Part C: Emerging Technologies, 52*, 74–92.
6. Stålhane, M., Halvorsen-Weare, E. E., Nonås, L. M., & Pantuso, G. (2019). Optimizing vessel fleet size and mix to support maintenance operations at offshore wind farms. *European Journal of Operational Research, 276*(2), 495–509.
7. Stålhane, M., Bolstad, K. H., Joshi, M., & Hvattum, L. M. (2021). A dual-level stochastic fleet size and mix problem for offshore wind farm maintenance operations. *INFOR: Information Systems and Operational Research, 59*(2), 1–33.
8. Irawan, C. A., Eskandarpour, M., Ouelhadj, D., & Jones, D. (2021). Simulation-based optimisation for stochastic maintenance routing in an offshore wind farm. *European Journal of Operational Research, 289*(3), 912–926.
9. Stock-Williams, C., & Swamy, S. K. (2019). Automated daily maintenance planning for offshore wind farms. *Renewable Energy, 133*, 1393–1403.
10. Dawid, R., McMillan, D., & Revie, M. (2016). *Development of an O&M tool for short term decision making applied to offshore wind farms*. WindEurope Summit 2016.
11. Li, X., Ouelhadj, D., Song, X., Jones, D., Wall, G., Howell, K. H., Igwe, P., Martin, S., Song, D., & Pertin, E. (2016). A decision support system for strategic maintenance planning in offshore wind farms. *Renewable Energy, 99*, 784–799, ISSN 0960-1481.
12. Li, M., Jiang, X., Carroll, J., & Negenborm, R. R. (2022). A multi-objective maintenance strategy optimization framework for offshore wind farms considering uncertainty. *Applied Energy, 321*, 119284.
13. Ren, Z., Verma, A. S., Li, Y., Teuwen, J. E., & Jiang, Z. (2021). Offshore wind turbine operations and maintenance: A state-of-the-art review. *Renewable and Sustainable Energy Reviews, 144*, 110886, ISSN 1364-0321.
14. Manupati, V. M., Schoenherr, T., Wagner, S. M., Soni, B., Panigrahi, S., & Ramkumar, M. (2021). Convalescent plasma bank facility location-allocation problem for COVID-19. *Transportation Research Part E: Logistics and Transportation Review, 156*, 102517, ISSN 1366-5545.
15. Campuzano, G., Obreque, C., & Aguayo, M. M. (2020). Accelerating the Miller–Tucker–Zemlin model for the asymmetric traveling salesman problem. *Expert Systems with Applications, 148*, 113229, ISSN 0957-4174.
16. Kallehauge, B., Larsen, J., Madsen, O. B., & Solomon, M. M. (2005). Vehicle routing problem with time windows. In G. Desaulniers, J. Desrosiers, & M. M. Solomon (Eds.), *Column generation*. Springer.
17. El-Sherbeny, N. A. (2010). *Vehicle routing with time windows: An overview of exact, heuristic and metaheuristic methods*. Mathematics Department, Faculty of Science, Al-Azhar University, Nasr City 11884, Cairo, Egypt.
18. World-Energy. (2022). *Wind power*. https://www.world-energy.org/article/27732.html
19. Dogger Bank. (2023). *Dogger bank wind farm*. https://doggerbank.com/

Privacy-Preserving Energy Trading with Applications to Renewable Energy Communities

Simona Ramos and Connor Mcmenamin

1 Introduction

Renewable energy communities (RECs) encounter a spectrum of challenges that span technological, socio-economic, and regulatory dimensions. However, their functionality is restricted by the availability of primary resources, such as solar and wind power. Moreover, these sources tend to exhibit variable renewable energy production, characterized by stochastic patterns. Technical solutions such as advanced energy storage technologies (e.g., batteries) often involve high costs associated with implementing, maintaining, and depreciating such technologies. As noted by MIT Technology Review [1], fluctuating solar and wind power require lots of energy storage, and while lithium-ion batteries seem like the obvious choice, they are far too expensive to play a major role [2].

Successful REC hinges on active member involvement in consumption/production, along with coordination and incentive alignment among participants. In addition, the socio-economic challenges in RECs often extend to achieving equitable access to renewable energy among the community members where participants benefit across diverse demographic and economic groups. Nevertheless, community owned production units and collectively owned renewable energy are often considered common goods, facing problems such as efficient allocation. This can often cause free-riding challenges such as over-consumption and/or under-provision of RE, underlining the well-known "Tragedy of the Commons" phenomena. Peer-to-peer (P2P) trading is often suggested as a solution to tackle challenges like free riding where members individually own production units and trade surplus energy in a market-based approach. However, there is a

S. Ramos (✉) and C. Mcmenamin
Universitat Pompeu Fabra, Barcelona, Spain
e-mail: simona.ramos@upf.edu

P. Pong (ed.), *Renewable Energy Resources and Conservation*, Green Energy and Technology, https://doi.org/10.1007/978-3-031-59005-4_12

possibility that these mechanisms may disproportionately favor members with greater resources, potentially leading to an imbalance within the community and, consequently, impacting the socio-economic fabric of an REC [3, 4]. Likewise, while P2P trading offers opportunities for individuals to directly exchange energy, it also raises questions about data privacy and confidentiality that need to be effectively addressed, particularly in line with the ongoing regulatory requirements [5, 6]. Further to this, RECs typically find themselves with a need to monitor and verify sustainable practices to avoid "greenwashing," and secure future funding, permits, and tax discounts.

Therefore, there is a clear need for a comprehensive and adaptable strategy, integrating technological innovation, effective coordination mechanisms, robust member data protection, and a reliable verification tools. By amalgamating these elements, sustainable change can be effectively propelled forward. In this chapter, we present a blockchain-based solution that achieves all of these needs. Specifically, we propose a solution (see Sect. 4) that achieves the following four important objectives:

1. Allow for the coordination of all community members, and increase the overall welfare of the community vs. individually rational strategies. We provide members with an incentive compatible mechanism[1] to trade energy in a single shared community marketplace before trading with the public energy grid. This ensures members both maximize community usage of renewable energy and retain of wealth within the community.
2. Enable privacy-preserving expression of supply and demand by default.
3. Generate and monitor green blockchain-based tokens that enable alignment of community goals with individual incentives and provide a monetizable medium with which we can ensure members follow our community welfare-optimizing protocol (see Sects. 3.1 and 5).
4. Enable relevant external parties (e.g., governments, organizations, municipality) to verify the sustainable behavior and progress of RECs/REC members (which can be often times crucial for granting funding, allowing to occupy land, reducing taxes, etc.).

Recognizing the inherent diversity within RECs, each marked by a unique organizational framework and specific challenges, this chapter's objective is not to offer an all-encompassing remedy for RECs. Rather, our focus centers on a targeted range of issues. For these, we intend to present viable solutions that can be refined and broadened to encompass a wider array of use case scenarios. Throughout the chapter, we employ the example of Culatra to elucidate and enhance reader comprehension. Culatra is a local renewable energy community at Culatra Island, located in the south of Portugal [2].

[1] Through our use of green tokens coupled with pricing that is, at worst, the same price achieved by trading directly with the public energy grid.

2 Related Work

The adoption of blockchain technology within RECs holds significant potential to enhance their operational framework, unlocking benefits that cannot exist in centralized and/or uncoordinated settings. Blockchain, characterized by its decentralized and transparent nature, offers a novel approach to addressing several critical challenges faced by RECs. An EU report maintains that blockchain can truly engage prosumers in the energy market acting as enabler for the creation of energy communities [7]. According to the study, blockchain enhances the transparency and trust of the energy market system.

Wang et al. [8], Wen et al. [9], and Guo et al. [10] develop blockchain-based solutions for some of the many challenges faced by RECs. The protocol in [8] leverages blockchain-based smart contracts to establish a P2P market for energy where local producers trade with local consumers, although overlooks the need to ensure user data remains private throughout the process. Wen et al. [9] propose a blockchain to enhance energy prices for demand-side management using demand response. The authors suggest the usage of pseudo-digital identity to enhance members' privacy. The authors do not go into further details explaining the design, implementation, and the management of these identities. Likewise, a recent review of blockchain-based energy trading platforms [10] identifies member privacy while also ensuring verifiability of the exchange process as an important open problem. This is something we solve in our protocol through our use of homomorphic encryption. The review also mentions compatibility for low-resource smart devices and scalability as important issues for blockchain-based energy trading. Low-resource individuals in our protocol (Sect. 4) are only required to verify their own data has been encrypted and decrypted correctly, as long as any one member in the community verifies the settlement price for each time slot has been performed correctly. Furthermore, as this verification is done locally (not computed using shared on-chain computation resources), this does not affect the scalability of the system.

Related to P2P trading as a means of increasing community welfare in micro-grids, [11–13] all introduce variations of P2P energy marketplaces enabling the exchange of energy between consumers and prosumers, leaning on the ability for users to set their own pricing mechanism. Compared to these, our solution is intended to align more with the socio-economic fabric of REC. We incentivize members to engage in energy trading as a unified batch through our use of green tokens, providing a clear optimal for both members and the community over any free-market approach. This is as a result of batch trading on its own maximizing community welfare through its optimal pricing guarantees [14], with green tokens dominating any potential benefit of free-market trading for the individual members. This not only promotes a sense of community collaboration, but also optimizes the efficiency of energy distribution within the REC, reinforcing its sustainable and interconnected nature. Kolahan et al. [15] argue for the implementation of smart controllers that estimate the probability of energy use in the next hour, in order to predict occupancy patterns and assist with demand management. The authors leverage blockchain-based network for buildings to exchange data of a specific parameter called the probability of the next hour. We allow for such control systems too, generalizing to notion of predictions based on historical data to any predictive mechanism, including the use of user-input information (not just extrapolating on historical data).

In terms of tokenomics, [12] design a blockchain-based asset ownership system allowing consumers to securely obtain energy production shares within a potential REC in Germany. Rozas et al. [16] suggest how blockchain-based affordances including tokenomics can theoretically be used to fulfill and automate Ostrom's principles, as a way toward avoiding the Tragedy of the Commons. Cila et al. [17] extend the debate arguing specific design dilemmas when creating a blockchain system, following a fictitious example of an energy community. These demonstrate some of the ways in which tokenized assets can play a role in the context of RECs by facilitating the representation and exchange of value within the community ecosystem. In our protocol, we leverage this value representation of token to incentivize the correct behavior of individually rational members in the community toward optimizing the overall welfare of the community.

3 Designing an Energy Marketplace for RECs

Creating a fair energy marketplace for RECs presents a multifaceted challenge. Often the main focus of these communities revolves around fostering socio-economic and environmental well-being rather than pursuing pure profit-maximization strategies [4]. As noted by Cutore et al. [18], there is often a trade-off between economic and social performances in RECs which should be addressed ex ante the implementation and design of the energy marketplace.

In this chapter, one of our key motivations is to uphold the community socio-economic fabric and motivating sustainable behaviors among its members, while providing a protocol toward maximizing the welfare of the community as a whole. The main challenge of any energy marketplace is the alignment of supply and demand. A free-market approach is a straightforward approach to solve this, although such an approach can be considered contrary to an REC's socio-economic fabric [19]. According to [20], dynamic pricing also requires sophisticated control technology and considerable implementation and operational costs and has not proven highly successful in RECs. Moreover, this approach risks generating ethical concerns if prices surge due to demand surpassing supply, potentially burdening vulnerable community members.

In a microgrid environment such as Culatra, where the microgrid is connected to the main grid, the community is typically given a price p_s/MWh to sell excess energy to the main grid, while receiving some price p_b/MWh to buy energy from the main grid, with $p_b > p_s$. From a recent analysis on Culatra, $p_s \approx$ €40/MWh [2], while current estimates for the buy price set $p_b \approx$ €100/MWh [21] (these numbers are used as indicative example). By creating a blockchain-based settlement process for matching supply and demand imbalances, we can use blockchain tokens to incentivize and describe collaborative behavior. One (simple) way we can use the tokens/nonmonetary incentives to incentivize cooperative behavior is:

- When supply exceeds demand, trade everything (sum of local net supply) at p_s, and give "green" tokens to the sellers that can be used to avail of community discounts and/or satisfy grant delivery conditions.
- When demand exceeds supply, trade everything (net demand) at p_b, and give the same "green" tokens to the buyers, again creating a dominant incentive to take part in the community settlement process.

In order to incentivize the community to act cooperatively and not individually, as well as to improve the supply and demand matching, we suggest that participating community members can submit their net energy usages (energy usage minus energy creation) at some point before each time slot using a homomorphic encryption scheme (explained in Sect. 4). Then, the auctioneer, a semi-trusted third party in charge of matching supply and demand, aggregates the individual net usages for each time slot. This aggregation takes the form of a single number representing the community's net usage for the respective time slot, without revealing individual usages. This net usage is then communicated with external sources, either purchasing energy from the main grid in the case of net demand or utilizing surplus energy to generate revenue by selling to the main grid (or alternative use cases [22]) in the case of net supply.

3.1 Green Tokens and Specific Use Cases

The drive for demand, value, and usage of green tokens is specific to the REC in which the tokens are deployed and typically depend on local conditions and needs. The following are examples of how demand can be created for the green tokens introduced:

(a) In line with [16]'s argument for effective decentralized governance of common goods, tokens can be used in REC to define community membership and voting rights.
(b) Tokens can be employed to provide local discounts on goods and services, encouraging community members to patronize local businesses and contribute to the growth of the local economy. This localized incentive mechanism not only bolsters community cohesion but also reinforces the REC's commitment to enhancing local sustainability and resilience.
(c) Green tokens can be used to satisfy national or international quotas for sustainable energy usage. By tying grants, such as land permits for occupying natural reserve land [23], to tangible assets, there is a clear value proposition for such tokens. This is in line with regulatory efforts to establish the universal and harmonized provisions for monitoring, reporting, and verification in climate change mitigation projects [24].

4 Protocol Description

This section outlines the blockchain protocol intended for a deployment in a potential REC community. In this section we first outline the model assumptions, and cryptographic primitives that are required for use in our protocol. We then merge these with the necessary blockchain functionalities, describing the entire protocol, as implemented here [25].

At a high level, our protocol implements a publicly verifiable and privacy-preserving supply–demand matching protocol. Participating community members submit their net energy usages (energy usage minus energy creation) at some point before each time slot using a homomorphic encryption scheme. The auctioneer, a special semi-trusted entity in the community in charge of matching supply and demand, aggregates the individual net usages for each time slot, outputting the community's overall supply or demand for the respective time slot, without revealing individual usages. An accurate prediction of energy demand within a microgrid can be crucial for ensuring an appropriate balance between supply and consumption [4]. Incentivizing members to report accurate energy forecasts can be used to secure better pricing in advance of such spikes, in the same way that energy producers trade futures on energy prices to minimize variance in profits. Important usage information that is typically known in advance could take the form of holiday plans for community members (reduced energy usage) versus increased demand for community hotels (increased energy usage). This, in conjunction with the ability to report these profiles in a privacy-preserving manner, has clear potential for an REC.

The net community usage is then communicated with external sources, either purchasing energy in the case of net demand, or utilizing surplus energy to generate revenue (selling to the grid, or alternative use cases like Bitcoin mining [22]) in the case of net supply. Through our choice of encryption scheme, the individual contributions to the net supply/demand can be communicated and recorded publicly without any individual's information leakage. This is done in a way that only requires the encrypted total for each community member at each time slot to be stored. All of this, while ensuring each encrypted total is valid with respect to the community total.

With a public record of community renewable energy usage at each time slot, and encrypted summary statistics for each individual, these individuals can verify to local, national, or international entities that certain quotas are being met. If individuals are responsible for such proofs, these same individuals can be sure that sensitive information leaked by granular energy usage statistics is avoided.

4.1 Model Assumptions

1. A public-key infrastructure exists such that for any public key, and a message encrypted using that public key, only the owner of the private key corresponding to the private key can decrypt the message.

2. There exists an auctioneer in our system who is trusted to keep his own private key and decrypted plaintext messages private.

Importantly, our model does not require any trust that the auctioneer performs the settlement process correctly. Through our choice of homomorphic encryption system, every member in the system can verify that the auctioneer is settling the auction correctly, and neither creates nor destroys wealth within the community.

4.1.1 Model Limitations

The trusted auctioneer becomes a single point of failure in this model. If the auctioneer becomes corrupted, all user usage profiles can be read. Precautions can be taken against this in practice, such as periodically rotating the auctioneer and/or the auctioneer's key pairs. This can restrict the fallout of attacks to only partial profile revelation. Decentralizing the auctioneer's role in this part of the protocol, such as through the use of threshold-key encryption or multiparty computation, makes for important future work.

4.2 Homomorphic Encryption Scheme

For the purposes of our protocol, we require an encryption scheme that for encrypted usages $e(u_i)$ of each member $i \in [1, \ldots, n]$, we can verify that for some value v, $v = \sum_{i=1}^{n} u_i$ without revealing any of the u_is. The Paillier encryption scheme is such a protocol [26]. For full details on the system, and requirements for key generation, see [27]. A Paillier private key can be described by a tuple (n, g, λ, μ), with corresponding public key (n, g). The encryption function e for $m \in [1, \ldots, n-1]$ the plaintext is described as

$$e : m, r \rightarrow r^n g^m \bmod n^2, \text{ for a random } r \text{ with } r, n \text{ coprime.} \tag{1}$$

The decryption function d for a ciphertext c is

$$d : c \rightarrow \lfloor \frac{c^\lambda \bmod n^2}{n} \rfloor . \mu \bmod n. \tag{2}$$

For $c = e(m, r)$, $d()$ is such that $d(e(m, r)) = m$. Importantly for our purposes, the Paillier encryption also has the following homomorphic property:

$$d(e(u_1).e(u_2)) = d(e(u_1)) + d(e(u_2)) = u_1 + u_2. \tag{3}$$

In our system, each member will post the tuple $(c_{i,r}, c_{i,m})$, an encryption of r_i and u_i, respectively, to the blockchain using the auctioneer's public key (n_a, g_a). Let these ciphertexts be $c_{i,r} = e(r_i, R_i)$ for some randomly chosen R_i, and $c_{i,m} = e(u_i, r_i)$, using the same r_i in both ciphertexts. This allows the auctioneer exclusively to decrypt each of the plaintexts and corresponding randomnesses (u_i, r_i). By posting $r = \prod_{i=1}^{n} r_i$ and $v = \sum_{i=1}^{n} u_i$, anyone can then verify that the sum is correct by checking $e(v, r) = \prod_{i=1}^{n} c_{i,m}$.[2]

4.3 Blockchain Protocol

Given these important functionalities, we are equipped to implement our blockchain protocol. In the blockchain protocol, each community individual is represented by an address, with the set of addresses controlled by a public-key infrastructure. The blockchain protocol progresses in real time. Unless otherwise specified, $e_i()/d_i()$ indicates Paillier encryption/decryption with member i's public/private key, while $e_a()/d_a()$ indicates Paillier encryption/decryption with the auctioneer's public/private key. Algorithm 1 describes the functions and data structures that are executed and stored on the blockchain. The protocol proceeds in sequential time slots, with each time slot requiring the following three steps:

1. **UPLOAD_ENC_USAGE()**: To be called by each member in the community for each time slot. The member decides on their net usage for the time slot u, some randomness r with which to encrypt that usage, and another randomness r' with which to encrypt the randomness r. These encryptions are done using the auctioneers public key, meaning only the auctioneer can decrypt the plaintext values. The encryption of r using r' allows the auctioneer to decrypt r, and as such compute the product of the randomnesses used to encrypt the usages. This product, along with the sum of the usages, allows any blockchain observer to verify that the encryptions and decryptions were all done correctly. Through calling **UPLOAD_ENC_USAGE()**, the member uploads the encryptions of u using r, and r using r', which are then stored on the blockchain in `enc_Usages` and `enc_Rands`, respectively. These values are to be used later by the auctioneer, and then by anyone wishing to verify the outcome of a particular auction.
2. **UPLOAD_DEC_USAGE()**: To be called by the auctioneer. After all members have called **UPLOAD_ENC_USAGE()** for a particular time slot, this blockchain function reveals the sum of the member usages for that time slot. Moreover, this function also reveals the product of the randomnesses used to encrypt each of these individual usages. For encrypted member usages $c_u^1, \ldots, c_u^n$ for a given time slot, with v the proposed net usage for the time slot, and r the proposed product of the randomnesses used, any blockchain member can verify that $\prod_{i=1}^{n} c_u^i = e_a(v, r)$. This ensures the decryption was done correctly.

[2] In order to compute $r = \prod_{i=1}^{n} r_i$, necessary for verification that a proposed v is indeed the sum of the individual plaintexts, these randomnesses must be encrypted separately. This is because decryption of a plaintext does not reveal the randomness used in encryption.

Algorithm 1 Blockchain protocol

```
1: mapping() enc_Usages
2: mapping() enc_Rands
3: mapping() dec_Net_Usage
4: mapping() dec_Rands_Product
5: mapping() enc_Tokens
6: mapping() enc_Token_Rands
7: function UPLOAD_ENC_USAGE(i=USER NUMBER,c_u=ENCRYPTED USAGE, c_r=ENCRYPTED
   RANDOMNESS, slot=TIME SLOT)
8:    enc_Usages[slot][i]=c_r
9:    enc_Usages[slot][i]=c_u
10: function UPLOAD_DEC_USAGE(v=TOTAL NET USAGE,r=RANDOMNESS PRODUCT USED, slot=TIME
    SLOT)
11:    dec_Net_Usage[slot]=v
12:    dec_Rands_Product[slot]=r
13: function UPDATE_USER_TOKENS(T=TOKEN UPDATES, R=ENCRYPTED RANDOMNESSES USED,
    slot=TIME SLOT, prod=PRODUCT OF RANDOMNESS, price= ENERGY PRICE USED)
14:    require(∏_{t∈T} t=e_a(price* dec_Net_Usage[slot], prod))
15:    for i ∈ [1, num_user] do
16:       enc_tokens_randomness[slot][i]=R[i]
17:       enc_tokens[i]=enc_tokens[i]*T[i]
```

3. **UPDATE_USER_TOKENS()**: To be called by the auctioneer to update the encrypted representation of a member's total tokens within the blockchain system, stored in `enc_tokens`. Each member can verify that their own encryption has been performed correctly, while also verifying that the sum of the encryptions matches the implied total from the decrypted total usage for the given time slot. As the encrypted token updates are done using $e_a()$, the blockchain performs a `require()` check before updating the token balances to ensure the token updates correspond to the net usage for that slot. Specifically, in line 14 the blockchain function checks that the decrypted net usage for the specified slot, `dec_Net_Usage[`$slot$`]`, times the price $price$, when encrypted with the specified randomness product $prod$ equals the product of the individual token updates, $\prod_{t \in T} t$. By the homomorphic property of the encryption scheme, this only holds true if the token updates equal the net usage multiplied by the specified energy price. Assuming each member verifies their own token update is done correctly, a reasonable assumption given members are token maximizing, the community as a whole can be confident that all token updates are performed correctly.

5 Tokens and Value Exchange Within an REC

This section introduces possible uses for such a green token (Sect. 3.1), mirroring the notion of carbon credits. We also describe how this notion can be enhanced by using the same privacy-preserving and verifiable functionalities of Sect. 4 to describe a marketplace for members to buy and sell these tokens without leaking sensitive information such as balances and trade history (Sect. 5).

The protocol described in Sect. 4 allows members to securely express supply and demand without revealing these preferences to other members. This supply and demand is converted into continuously updated and verifiable homomorphically encrypted financial balances for each member. The intention with such balances is to, at a minimum, track the amount owed to or by each community member. This core protocol implicitly records each individual's participation rates, self-sufficiency ratio, and self-consumption ratio. All of these variables can be explicitly recorded (in a privacy-preserving way if necessary) and merged to translate our high-level proposal of green tokens into a more tangible value proposition as described in Sect. 3.1.

Regardless of the exact use case for the green tokens, there are many uses that create utility for users that can be translated to monetary value. To motivate the value proposition of these tokens, consider the use of green tokens for use in community improvement proposals and voting. In renewable energy communities such as Culatra, there are numerous shared costs related to infrastructure and development that must be prioritized. Green tokens received for participation in our proposed welfare-maximizing protocol of Sect. 4 are ideal for this purpose. Green tokens in such a system should then be distributed proportionally to volume traded, although in line with the democratic needs of the community. Specifically, it is likely important to prevent monopolization of green tokens, so users may need to be pro-rated based on their expected volume/usage, while still incentivizing maximal volume to trade within the community protocol. Tokens can then be destroyed (anonymously, again using the same techniques as introduced in Sect. 4.2), and exchanged for votes. With competing utilities for one vote over another from each member's perspective,[3] tokens now have a monetizable value with users having clear motivation to buy and sell. Specifically, members can express the value of such a green token vote in monetary terms.

To allow for the exchange of such tokens, consider a community progressing through time with members accruing various quantities of these green tokens, while votes are periodically taking place. As discussed, there will be a natural desire for members to exchange tokens. With blockchain technology, there are many ways to implement such an exchange in a decentralized manner [28–30]. These protocols match buyers with sellers, implementing variations of a frequent batch auction [14]. Batch auctions involving a trusted auctioneer and encrypted order information, as in [28, 29], are proven to settle orders at a price representative of the true underlying supply and demand. As blockchain members can observe all of the auction inputs and outputs, we can again leverage homomorphic encryption to ensure the auction is settled correctly. This stands as another example of blockchain-based techniques that can enhance these renewable energy communities.

[3] Members living away from a set of proposed development sites for new wind turbines may be indifferent to the development location compared to members closer to some sites than others due to noise pollution.

6 Conclusion

We present a comprehensive framework that not only enhances the coordination, privacy, and alignment of incentives within RECs but also empowers them to establish a trustworthy reputation in the eyes of relevant external stakeholders. In particular, we introduce a privacy-preserving energy trading protocol that enables REC members to securely communicate their energy supply and demand. Coupled with our use of tokenized incentives, users are encouraged to publish usage profiles and trade energy in a community-controlled public forum. This allows all users in the community to benefit from typically cheaper locally produced renewable energy, while also allowing the community as a whole to more effectively to balance energy supply and demand. All of this is provided without compromising the confidentiality of sensitive financial and usage data of REC members. Through these advancements, our blockchain-based protocol contributes to the advancement of sustainable energy adoption at the community level and paves the way for broader societal and environmental benefits.

References

1. Technology Review (2018). *The $2.5 Trillion Reason We Can't Rely on Batteries to Clean Up the Grid.*
2. Pacheco, A., Monteiro, J., Santos, J., Sequeira, C., & Nunes J. (2022). Energy transition process and community engagement on geographic Islands: The case of Culatra Island (Ria Formosa, Portugal). *Renewable Energy, 184*, 700–711.
3. Junlakarn, S., Kokchang, P., & Audomvongseree, K. (2022). Drivers and challenges of peer-to-peer energy trading development in Thailand. *Energies, 15*, 1229.
4. Gui, E. (2019). *Investment Planning and Institution Design for Community Microgrids as a Socio-technical Energy System.*
5. de Almeida, L., & Klausmann, N. (2021). Peer-to-Peer energy communities: Legal definitions and access to markets. Journal Name.
6. Schneiders, A., & Shipworth, D. (2021). Community energy groups: Can they shield consumers from the risks of using blockchain for peer-to-peer energy trading? *Energies, 14*(12), 3569.
7. Commission, E. (2017). *Blockchain in energy communities.*
8. Wang, X., Yang, W., Noor, S., Chen, C., Guo, M., & van Dam, K. H. (2019). Blockchain-based smart contract for energy demand management. *Energy Procedia, 158*, 2719–2724.
9. Wen, S., Xiong, W., Tan, J., Chen, S., & Li, Q. (2021). Blockchain enhanced price incentive demand response for building user energy network in sustainable society. *Energy Reports, 7*, 2704–2712.
10. Guo, Y., Wan, Z., & Cheng, X. (2022). When blockchain meets smart grids: A comprehensive survey. *High-Confidence Computing, 2*(2), 100059.
11. Mengelkamp, E, Gärttner, J., Rock, K., Kessler, S., Orsini, L., & Weinhardt, C. (2018). Designing microgrid energy markets: A case study—The Brooklyn Microgrid. *Applied Energy, 210*, 870–880.
12. Durillon, B., Davigny, A., Kazmierczak, S., Barry, H., Saudemont, C., & Robyns, B. (2020). Decentralized neighbourhood energy management considering residential profiles and welfare for grid load smoothing. *Sustainable Cities and Society, 63*, 102464.

13. Noor, S., Yang, W., Guo, M., van Dam, K. H., & Wang, X. (2018). Energy demand side management within micro-grid networks enhanced by blockchain. *Applied Energy, 226*, 47–60.
14. Budish, E., Cramton, P., & Shim, J. (2015). The high-frequency trading arms race: Frequent batch auctions as a market design response *. *The Quarterly Journal of Economics 07, 130*(4), 1547–1621.
15. Kolahan, A., Maadi, S. R., Teymouri, Z., & Schenone C. (2021). Blockchain-based solution for energy demand-side management of residential buildings. *Energy Reports, 7*, 1813–1824.
16. Rozas, D., Tenorio-Fornés, A., Díaz-Molina, S., & Hassan, S. (2021). When Ostrom Meets Blockchain: Exploring the Potentials of Blockchain for Commons Governance. *SAGE Open, 11*(1), 21582440211002526.
17. Cila, N., Ferri, G., de Waal, M., Gloerich, I., & Karpinski, T. (2020). *The blockchain and the commons: Dilemmas in the design of local platforms*.
18. Cutore, E., Volpe, R., Sgroi, R., & Fichera, A. (2023). Energy management and sustainability assessment of renewable energy communities: The Italian context. *Energy Conversion and Management, 278*, 116713.
19. Gan, L., Jiang, P., Lev, B., & Zhou, X. (2020). Balancing of supply and demand of renewable energy power system: A review and bibliometric analysis. *Sustainable Futures, 2*, 100013.
20. Siano, P. (2014). Demand response and smart grids–A survey. *Renewable and Sustainable Energy Reviews, 30*, 461–478.
21. OMIE (2023). Spanish and Portuguese Energy Prices. Accessed: 14/08/2023. https://www.omie.es/pt.
22. Scilling (2023). Digital Mining. Accessed: 14/08/2023. https://www.scillingmining.com/.
23. Rees, P. (2017). Culatra Island home-owners to be granted 30-year residence licences. Algarve Daily News.
24. Singh, N., Finnegan, J., Levin, K., Damassa, T., Elsayed, S., Mitra, A., et al. (2016). *Understanding measurement, reporting, and verification of climate change mitigation*.
25. Github (2023). GitHub. https://github.com/The-CTra1n/RE-Communities.
26. Paillier, P. (2005). *Paillier encryption and signature schemes*.
27. Will, M. A., & Ko R. K. L. (2015). Chapter 5—A guide to homomorphic encryption. In R. Ko, K. K. R. Choo (Eds.), *The cloud security ecosystem* (pp. 101–127). Boston: Syngress.
28. Penumbra. Accessed: 23/07/2023. https://penumbra.zone/.
29. McMenamin, C., Daza, V., Fitzi, M., & O'Donoghue, P. (2022). FairTraDEX: A decentralised exchange preventing value extraction. In *Proceedings of the 2022 ACM CCS workshop on decentralized finance and security (DeFi'22)* (pp. 39–46). New York: Association for Computing Machinery.
30. CoW Protocol. Accessed: 11/08/2023. https://docs.cow.fi/.

Use of Watermelon Waste As a Fuel Source for Bioelectricity Generation

Rojas-Flores Segundo, Santiago M. Benites, De La Cruz-Noriega Magaly, Nazario-Naveda Renny, Nélida Milly Otiniano, and Daniel Delfín-Narciso

1 Introduction

Organic waste has become a big problem for human society, mainly due to the exponential increase in the production of fruits and vegetables by agro-industrial companies, which, due to increasing demand, have been forced to increase their production without seeing the consequences [1–3]. According to the United Nations World Organization, it has been estimated that people throw away approximately one third (1.6 billion tons) of the food produced each year, of which 40–50% represent vegetable and fruit waste [4, 5]. This opens a new area of research that has to be covered urgently, since an increase of 30% is estimated for the year 2030. Solving this problem would benefit companies and society [6]. Currently, fruit waste is being used in various ways, for example, for the generation of biogas, compost, fertilizers, jams, cosmetics, bioelectricity, etc. [7, 8]. Generation of bioelectricity is one of the least addressed fields where a greater amount of research is needed [9]. One of the ways to generate electrical energy from fruit waste is through the use of microbial fuel cells (MFCs) where waste is used as fuel and through redox reactions that occur within these electronic devices, electrons are produced that when flow

R.-F. Segundo (✉) · S. M. Benites · D. L. C.-N. Magaly · N.-N. Renny
Vicerrectorado de Investigación, Universidad Autónoma del Perú, Lima, Peru
e-mail: santiago.benites@autonoma.pe; mdelacruzn@autonoma.edu.pe; renny.nazario@autonoma.pe

N. M. Otiniano
Instituto de Investigación en Ciencias y Tecnología de la Universidad Cesar Vallejo, Trujillo, Peru
e-mail: notiniano@ucv.edu.pe

D. Delfín-Narciso
Grupo de Investigación en Ciencias Aplicadas y Nuevas Tecnologías, Universidad Privada del Norte, Trujillo, Peru
e-mail: daniel.delfin@upn.edu.pe

P. Pong (ed.), *Renewable Energy Resources and Conservation*, Green Energy and Technology, https://doi.org/10.1007/978-3-031-59005-4_13

within a circuit generate electric current [10–12]. One of the most widely used MFC designs are single-chamber microbial fuel cells (MFC-SC) due to their low manufacturing cost and high current and voltage values shown in previous works [13, 14].

On the other hand, watermelon waste at this time of year in Peru (South America) and in a large number of countries with a tropical climate is being intensely produced and consumed, where the pulp, seeds, and juice of the watermelon is used for consumption. while the shell, which represents 30% of the product, is discarded without any known use [15, 16]. The fruit contains micronutrients (H, C, Na, N, Mg, K, P, and C) and macronutrients (sugar, lipids, and protein) which are beneficial for health, apart from being refreshing for people [17, 18]. It has been reported that this fruit in the year 2021 harvested approximately 101,620,420 tons of which approximately 21% have not been marketed because the product was damaged by various factors [19]. In this sense, research on the use of fruit waste in MFC in various forms has been reported, for example, Zafar et al. (2023) managed to generate 500 mV and 221 mW/m^2 peak voltages and power densities using apple waste from a local fruit market as substrate (fuel) in their single-chamber MFCs [20]. Whereas, Verma, M. and Mishra V. (2023) managed to generate voltage peaks and power density of 481.33 ± 3.51 mV and 204.80 ± 1.28 mW/m^2 using lemon peel as fuel in their double-cell microbial fuel cells, H-type chamber [21]. The waste from Waste Potatoes has also been used in MFC-SC as a substrate for the generation of electrical energy, managing to show a peak of 1.58 V and 0.3714 mW/cm^2 as voltage and maximum power density using carbon felts with ZnO nanoparticles as electrodes [22].

Due to this, this research has as main objective to demonstrate the potential of watermelon waste for the generation of bioelectricity through single-chamber microbial fuel cells on a laboratory scale. For this, the values of electric current, voltage, pH, and electrical conductivity will be monitored for 30 days. This will also calculate the internal resistance, current density, and power density of the MFC-SC. In this way, a value will be given to watermelon waste to be used as fuel, benefiting companies dedicated to importing this fruit, and society itself because they will be able to use their waste to generate environmentally friendly electricity. environment.

2 Materials and Methods

(a) Single chamber fuel cell design: Three single-chamber microbial fuel cells were manufactured, which were purchased from SAIDKOCC (100 mL volume, Fujian, China), where copper electrodes (Cu, area = 40 cm^2) were used for the anode and zinc for the cathode (Zn, area = 62.5 cm^2), which were connected on the outside by an external circuit (whose external resistance was 100 Ω). Nafion 117 (Merck) was used as a proton exchange membrane.
(b) Watermelon waste collection: The fuel used in the MFC-SC was watermelon waste collected from the Palermo Ex Mayorista market, Trujillo, Peru; 1.5 kg of

waste was collected, which was taken to the laboratory to be washed three times to eliminate any type of impurity acquired from the environment. The waste was crushed in an extractor (Labtron, LDO-B10-USA) obtaining approximately 800 mL.

(c) Characterization of microbial fuel cells: Current and voltage values were monitored for 30 days using a multimeter (Prasek Premium PR-85, USA), while power density and current density values were measured by the described method by Rojas-Flores et al. (2022) whose external resistances were the same as those used in the previous work [23], while the internal resistance was found using the energy sensor (Vernier- ±30 V & ±1000 mA, USA) and using ohm's law. Likewise, the pH and electrical conductivity values were monitored with a pH-meter (110 Series Oakton, USA) and a conductivity meter (CD-4301, USA), during the 30 days of operation.

3 Results and Analysis

Figure 1a shows the voltage values observed during the monitoring period, being able to observe that the values increase from the first day (0.192 ± 0.011 V) to day 14 (0.983 ± 0.059 V) and then slowly decrease until the last day (0.451 ± 0.064 V). The high voltage values obtained are due to the potential difference generated between the electrodes of the microbial fuel cells. This is because of the reduction that occurs in the cathode electrode due to bacterial catholic activity [24], while the decrease of these values observed in the final part of the monitoring may be due to the degradation of the zinc electrode used as testing, due to the chemical processes that occurred in the power generation process [25, 26]. Likewise, these values compared to those of the literature showed to be higher, for example, Priya A. and Setty Y. (2019) managed to generate voltage peaks of approximately 0.4 V using apple waste (300 mL) as fuel in their microbial fuel cells fabricated with carbon electrodes, which operated down to pH 4.1 [27]. This was also observed in the research carried out by Kondaveeti et al. (2019) where I was able to generate 219 mV peaks using citrus peel as fuel in their MFC-SC using graphite electrodes [28].

Figure 1b shows the values of electric current generated by the MFC-SC, where the values increased from the first day (0.405 ± 0.040 mA) to day 15 (4.575 ± 0.647 mA) and then decreased slightly until the last day (2.073 ± 0.820 mA). The rapid increase shown in the monitoring of the electric current values, in the first days, is due to the good formation of the biofilm on the anode electrode that, due to microbial activity, degrades the compounds present in the substrate, while the decrease in the last days of monitoring must be due to the scarcity of these components used as nutrients for the metabolism of microorganisms that generate electrical energy [29–31]. Figure 1c shows the monitored pH values, where the values were maintained in the slightly acid regime, with an optimum operating pH of 5.84 ± 0.27 on the fifteenth day. All microbial fuel cells have their own optimum operating pH values, which depend on the substrate

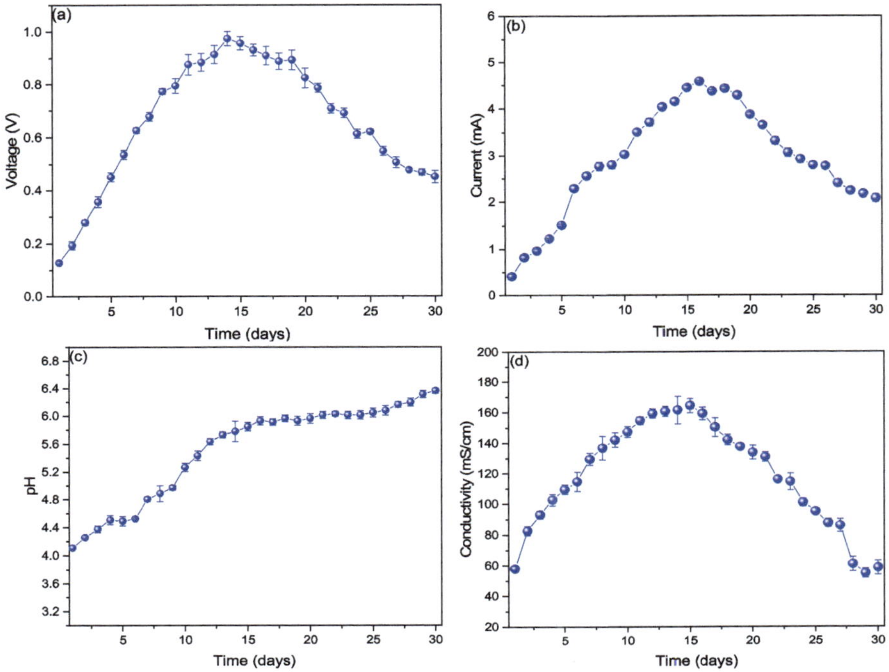

Fig. 1 Values obtained from (**a**) voltage, (**b**) electrical current, (**c**) pH, and (**d**) electrical conductivity obtained from monitoring microbial fuel cells

used and how much electrical energy the MFCs will generate because electrical energy-producing microorganisms metabolize at specific pHs [32, 33]. Figure 1d shows the values of electrical conductivity, observing an increase from the first day (57.79 ± 1.73 mS/cm) to the fifteenth day (164.87 ± 0.65 mS/cm) until the last day (58.56 ± 4.51 mS/cm). The electrical conductivity values increased in the first days due to the high ionic content present in the waste initially, but as the days passed they decreased due to the degradation or fermentation of the substrate [34].

The internal resistance was calculated using Ohm's Law, $V = IR$, where the voltage values were placed on the "x" axis and the electric current values on the "y" axis, which by linear adjustment would represent the slope. the internal resistance of single-chamber microbial fuel cells (see Fig. 2a). The calculated internal resistance was 36.748 ± 2.747 Ω; this value shows the high conductivity that exists in this experiment and the low resistance to the passage of electrons throughout the circuit. This would be due to the metallic nature of the electrodes used and good formation of the anodic biofilm mainly [35, 36]. Figure 2b shows the power density (PD) values as a function of current density (CD); calculating the maximum PD was 754 mW/m^2 at a CD of 4.51 mA/cm^2 with a peak voltage of 872.97 ± 21.51 V. Latif et al. (2020) managed to generate a maximum power density of 62 mW/m^2 at a current density of 229 mA/m^2 using debris from pineapples, oranges, bananas,

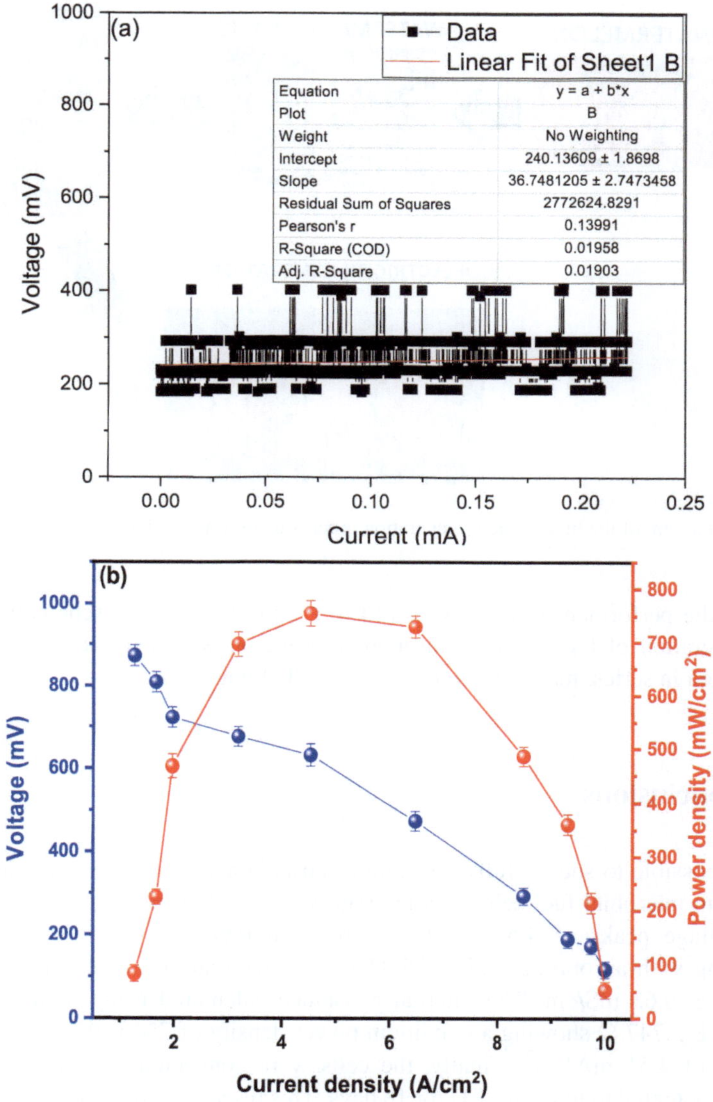

Fig. 2 (**a**) Internal resistance and (**b**) power density as a function of current density of microbial fuel cells

watermelons, mangoes, and papayas as a substrate [37]. While Asefi et al. (2019) managed to generate a voltage and power density of 775 ± 21 mV and 422 mW/m^2 at a current density of 850 mA/cm^2 using food waste as a substrate using carbon felts as electrodes [38]. Likewise, Ghazali et al. (2019) mention in their research that the use of biocatalysts and the reduction of the distance between electrodes influence the generation of power density, as well as the use of electrodes sensitive to humidity

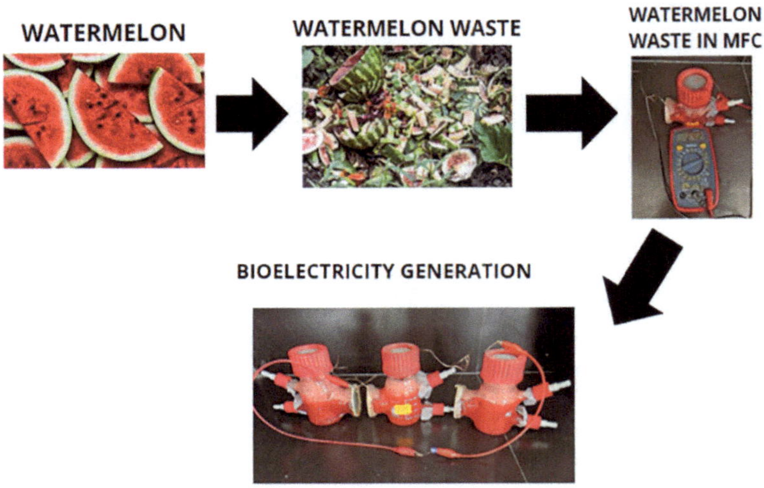

Fig. 3 Diagram of the bioelectricity generation process using watermelon waste as fuel

impair the performance of the MFCs [39]. Figure 3 shows the schematization of the generation of bioelectricity through watermelon waste, where the cells were connected in series, managing to turn on an LED light.

4 Conclusions

It was possible to successfully generate electrical energy using watermelon waste as fuel in microbial fuel cells on a laboratory scale. Managing to generate current and voltage peaks of 4.575 ± 0.647 mA and 0.983 ± 0.059 V on day 15, operating with an optimum pH of 5.84 ± 0.27 with an electrical conductivity of 164.87 ± 0.65 mS/cm. The internal resistance calculated using ohm's law was 36.748 ± 2.747 Ω showing a maximum power density of 754 mW/m^2 at a current density of 4.51 mA/cm^2. Finally, the cells were connected in series generating enough potential to light an LED for 18 days. This research shows the great potential of this type of waste for use by companies and farmers so that in the near future, when it is scalable, they can use their own waste to generate electricity.

References

1. Yaashikaa, P. R., Kumar, P. S., & Varjani, S. (2022). Valorization of agro-industrial wastes for biorefinery process and circular bioeconomy: A critical review. *Bioresource Technology, 343*, 126126.

2. Cremonez, P. A., Teleken, J. G., Meier, T. R. W., & Alves, H. J. (2021). Two-stage anaerobic digestion in agroindustrial waste treatment: A review. *Journal of Environmental Management, 281*, 111854.
3. Gaur, V. K., Sharma, P., Sirohi, R., Varjani, S., Taherzadeh, M. J., Chang, J. S., et al. (2022). Production of biosurfactants from agro-industrial waste and waste cooking oil in a circular bioeconomy: An overview. *Bioresource Technology, 343*, 126059.
4. Freitas, L. C., Barbosa, J. R., da Costa, A. L. C., Bezerra, F. W. F., Pinto, R. H. H., & de Carvalho Junior, R. N. (2021). From waste to sustainable industry: How can agro-industrial wastes help in the development of new products? *Resources, Conservation and Recycling, 169*, 105466.
5. Singh, R. S., Kaur, N., & Kennedy, J. F. (2019). Pullulan production from agro-industrial waste and its applications in food industry: A review. *Carbohydrate Polymers, 217*, 46–57.
6. Singh, R., Das, R., Sangwan, S., Rohatgi, B., Khanam, R., Peera, S. P. G., et al. (2021). Utilisation of agro-industrial waste for sustainable green production: A review. *Environmental Sustainability, 4*(4), 619–636.
7. Maddalwar, S., Nayak, K. K., Kumar, M., & Singh, L. (2021). Plant microbial fuel cell: Opportunities, challenges, and prospects. *Bioresource Technology, 341*, 125772.
8. Boas, J. V., Oliveira, V. B., Simões, M., & Pinto, A. M. (2022). Review on microbial fuel cells applications, developments and costs. *Journal of Environmental Management, 307*, 114525.
9. Palanisamy, G., Jung, H. Y., Sadhasivam, T., Kurkuri, M. D., Kim, S. C., & Roh, S. H. (2019). A comprehensive review on microbial fuel cell technologies: Processes, utilization, and advanced developments in electrodes and membranes. *Journal of Cleaner Production, 221*, 598–621.
10. Ramya, M., & Kumar, P. S. (2022). A review on recent advancements in bioenergy production using microbial fuel cells. *Chemosphere, 288*, 132512.
11. Yaqoob, A. A., Ibrahim, M. N. M., & Guerrero-Barajas, C. (2021). Modern trend of anodes in microbial fuel cells (MFCs): An overview. *Environmental Technology & Innovation, 23*, 101579.
12. Gul, H., Raza, W., Lee, J., Azam, M., Ashraf, M., & Kim, K. H. (2021). Progress in microbial fuel cell technology for wastewater treatment and energy harvesting. *Chemosphere, 281*, 130828.
13. Naseer, M. N., Zaidi, A. A., Khan, H., Kumar, S., Bin Owais, M. T., Jaafar, J., et al. (2021). Mapping the field of microbial fuel cell: A quantitative literature review (1970–2020). *Energy Reports, 7*, 4126–4138.
14. Jatoi, A. S., Akhter, F., Mazari, S. A., Sabzoi, N., Aziz, S., Soomro, S. A., et al. (2021). Advanced microbial fuel cell for waste water treatment—A review. *Environmental Science and Pollution Research, 28*, 5005–5019.
15. Letechipia, J. O., González-Trinidad, J., Júnez-Ferreira, H. E., Bautista-Capetillo, C., Rovelo, C. O. R., & Rodríguez, A. R. C. (2023). Removal of arsenic from semiarid area groundwater using a biosorbent from watermelon peel waste. *Heliyon, 9*(2), e13251.
16. Odewunmi, N. A., Umoren, S. A., & Gasem, Z. M. (2015). Watermelon waste products as green corrosion inhibitors for mild steel in HCl solution. *Journal of Environmental Chemical Engineering, 3*(1), 286–296.
17. Liu, C., Ngo, H. H., & Guo, W. (2012). Watermelon rind: Agro-waste or superior biosorbent? *Applied Biochemistry and Biotechnology, 167*, 1699–1715.
18. Hasanin, M. S., & Hashem, A. H. (2020). Eco-friendly, economic fungal universal medium from watermelon peel waste. *Journal of Microbiological Methods, 168*, 105802.
19. Yuan, Z., Sun, X., Hua, J., Zhu, Y., Yuan, J., & Qiu, F. (2023). Upcycling watermelon peel waste into a sustainable environment-friendly biochar for assessment of effective adsorption property. *Arabian Journal for Science and Engineering, 48*, 1–11.
20. Zafar, H., Peleato, N., & Roberts, D. (2023). A comparison of reactor configuration using a fruit waste fed two-stage anaerobic up-flow leachate reactor microbial fuel cell and a single-stage microbial fuel cell. *Bioresource Technology, 374*, 128778.
21. Verma, M., & Mishra, V. (2023). Bioelectricity generation using sweet lemon peels as Anolyte and cow urine as Catholyte in a yeast-based microbial fuel cell. *Waste and Biomass Valorization, 14*, 1–15.

22. Din, M. I., Ahmed, M., Ahmad, M., Iqbal, M., Ahmad, Z., Hussain, Z., et al. (2023). Investigating the activity of carbon fiber electrode for electricity generation from waste potatoes in a single-chambered microbial fuel cell. *Journal of Chemistry, 2023*, 8520657.
23. Rojas-Flores, S., Nazario-Naveda, R., Benites, S. M., Gallozzo-Cardenas, M., Delfín-Narciso, D., & Díaz, F. (2022). Use of pineapple waste as fuel in microbial fuel cell for the generation of bioelectricity. *Molecules, 27*(21), 7389.
24. Abazarian, E., Gheshlaghi, R., & Mahdavi, M. A. (2023). Interactions between sediment microbial fuel cells and voltage loss in series connection in open channels. *Fuel, 332*, 126028.
25. Jung, S. P., Son, S., & Koo, B. (2023). Reproducible polarization test methods and fair evaluation of polarization data by using interconversion factors in a single chamber cubic microbial fuel cell with a brush anode. *Journal of Cleaner Production, 390*, 136157.
26. Yaqoob, A. A., Fadzli, F. S., Ibrahim, M. N. M., & Yaakop, A. S. (2023). Benthic microbial fuel cells: A sustainable approach for metal remediation and electricity generation from sapodilla waste. *International journal of Environmental Science and Technology, 20*(4), 3927–3940.
27. Priya, A. D., & Setty, Y. P. (2019). Cashew apple juice as substrate for microbial fuel cell. *Fuel, 246*, 75–78.
28. Kondaveeti, S., Mohanakrishna, G., Kumar, A., Lai, C., Lee, J. K., & Kalia, V. C. (2019). Exploitation of citrus peel extract as a feedstock for power generation in microbial fuel cell (MFC). *Indian Journal of Microbiology, 59*, 476–481.
29. Mohyudin, S., Farooq, R., Jubeen, F., Rasheed, T., Fatima, M., & Sher, F. (2022). Microbial fuel cells a state-of-the-art technology for wastewater treatment and bioelectricity generation. *Environmental Research, 204*, 112387.
30. Mukherjee, A., Patel, V., Shah, M. T., Jadhav, D. A., Munshi, N. S., Chendake, A. D., & Pant, D. (2022). Effective power management system in stacked microbial fuel cells for onsite applications. *Journal of Power Sources, 517*, 230684.
31. Sharma, R., Kumari, R., Pant, D., & Malaviya, P. (2022). Bioelectricity generation from human urine and simultaneous nutrient recovery: Role of microbial fuel cells. *Chemosphere, 292*, 133437.
32. Wang, H., Wei, L., & Shen, J. (2022). Metal-free catalyst for efficient pH-universal oxygen reduction electrocatalysis in microbial fuel cell. *Journal of Electroanalytical Chemistry, 911*, 116233.
33. Littfinski, T., Beckmann, J., Gehring, T., Stricker, M., Nettmann, E., Krimmler, S., et al. (2022). Model-based identification of biological and pH gradient driven removal pathways of total ammonia nitrogen in single-chamber microbial fuel cells. *Chemical Engineering Journal, 431*, 133987.
34. Segundo, R. F., Magaly, D. L. C. N., Benites, S. M., Daniel, D. N., Angelats-Silva, L., Díaz, F., & Luis, C. C. (2022). Generation of electricity through papaya waste at different pH. *Environmental Research, Engineering & Management, 78*(4), 137.
35. López Zavala, M. Á., & Cámara Gutiérrez, I. C. (2023). Effects of external resistance, new electrode material, and Catholyte type on the energy generation and performance of dual-chamber microbial fuel cells. *Fermentation, 9*(4), 344.
36. Alvarez-Benítez, L., Silva-Martínez, S., Hernandez-Perez, A., Kamaraj, S. K., Abbas, S. Z., & Alvarez-Gallegos, A. (2022). Quantification of internal resistance contributions of sediment microbial fuel cells using petroleum-contaminated sediment enriched with kerosene. *Catalysts, 12*(8), 871.
37. Latif, M., Fajri, A. D., & Muharam, M. (2020). Penerapan sampah buah tropis untuk microbial fuel cell. *Journal Rekayasa Elektrika, 16*(1), 1.
38. Asefi, B., Li, S. L., Moreno, H. A., Sanchez-Torres, V., Hu, A., Li, J., & Yu, C. P. (2019). Characterization of electricity production and microbial community of food waste-fed microbial fuel cells. *Process Safety and Environmental Protection, 125*, 83–91.
39. Ghazali, N. F., Mahmood, N. A. N., Bakar, N. F. A., & Asri'Ibrahim, K. (2019). Temperature dependence of power generation of empty fruit bunch (EFB) based microbial fuel cell. *Malaysian Journal of Fundamental and Applied Sciences, 15*(4), 489–491.

Scenario Analysis on Deployment of Clean Liquid Fuels in Japan Toward Decarbonizing Energy Systems

Akito Ozawa, Yuki Kudoh, and Ruth Anne Gonocruz

1 Introduction

As concerns about climate change deepen, actions to achieve carbon neutrality have been taken around the world. Working Group 3 of the Intergovernmental Panel on Climate Change (IPCC) reported in the Sixth Assessment Report (AR6 WG3) [1] that global carbon dioxide (CO_2) emissions must be reduced to net zero by the early 2050s to limit global temperature increase to 1.5 °C above pre-industrial level. Carbon neutrality has become the most important global challenge, and many countries have set carbon neutrality as a long-term goal.

Japan has declared a goal of reducing its greenhouse gas (GHG) emissions by 46% compared to 2013 levels by FY2030 and achieving net zero by 2050. Achieving this ambitious GHG reduction targets requires a drastic energy transition. Japan's Strategic Energy Plan [2] includes the following initiatives to achieve carbon neutrality: (i) expansion of low-carbon power sources (renewable energy, nuclear power, thermal power with carbon capture and storage (CCS), hydrogen) and electrification; (ii) use of clean fuels in sectors that are difficult to electrify (hydrogen, e-fuel, biofuel, etc.); and (iii) introduction of carbon removal (DACCS, BECCS, forest absorption, etc.).

To implement a strategy toward carbon neutrality, policymakers must manage uncertainties regarding future technological developments and socioeconomic conditions. Several countries have applied scenario approaches to make strategic decisions under uncertainties. The Japanese government has also pointed out the need for this approach to explore pathways toward carbon neutrality [2].

A. Ozawa (✉) · Y. Kudoh · R. A. Gonocruz
Global Zero Emission Research Center (GZR), National Institute of Advanced Industrial Science and Technology (AIST), Tsukuba, Ibaraki, Japan
e-mail: akito.ozawa@aist.go.jp

P. Pong (ed.), *Renewable Energy Resources and Conservation*, Green Energy and Technology, https://doi.org/10.1007/978-3-031-59005-4_14

Against this background, mathematical model simulations have been conducted to clarify the energy transition pathways to achieve Japan's environmental target. Ozawa et al. [3, 4] investigated the role of hydrogen energy and carbon removal using an energy model. Sugiyama et al. [5] analyzed the scenarios to achieve Japan's CO_2 mitigation target by 2050 using five different energy models. Some studies assessed the ripple effects of energy transition in Japan, such as job creation [6, 7] and material requirements [8, 9].

All technological options must be investigated to explore pathways toward carbon neutrality under future uncertainties. This chapter presents a carbon-neutral scenario using clean liquid fuels for the transportation sector. Clean liquid fuels are considered important for reducing CO_2 emissions from hard-to-abate transportation and industry sectors. However, their role in Japan's carbon-neutral scenario has not been examined in previous studies. This study incorporated e-fuel and biofuel supply chains into the model to analyze Japan's energy transition toward net-zero CO_2 emissions by 2050. The simulation results were used to evaluate the transformations of the energy supply and technology deployment under different conditions for clean liquid fuels.

2 Methods

MARKAL is a mathematical model for energy system analysis developed in a project by the Energy Technology Systems Analysis Program of the International Energy Agency (IEA) [10]. By solving a linear programming problem, MARKAL identifies a combination of energy carriers and technologies that minimizes the total energy system cost Z, which is defined by Eq. (1),

$$Z = \sum_{r=1}^{R} \sum_{t=1}^{\mathrm{NPER}} \mathrm{ANNCOST}(r,t) \cdot \Big((1+d)^{-\mathrm{NYRS}\cdot(t-1)} + (1+d)^{-\mathrm{NYRS}\cdot(t-1)-1} + (1+d)^{-\mathrm{NYRS}\cdot(t-1)-2} + \cdots + (1+d)^{-\mathrm{NYRS}\cdot t+1} \Big) \quad (1)$$

where ANNCOST is the annual cost, which is the sum of construction and operation costs for energy technologies, and mining, import, transport, and distribution costs for energy carriers; r is regions; t is periods; d is the discount rate; NPER is the number of periods; and NYRS is the number of years in each period. The MARKAL model also requires that constraints on the energy system, including technology capacity transfer, energy balance, and CO_2 reduction targets, be met. Further descriptions of the MARKAL model are provided in Loulou et al. [11].

This study used a MARKAL model developed by the AIST, for energy system analysis in Japan. The AIST-MARKAL model considers Japan as a single region

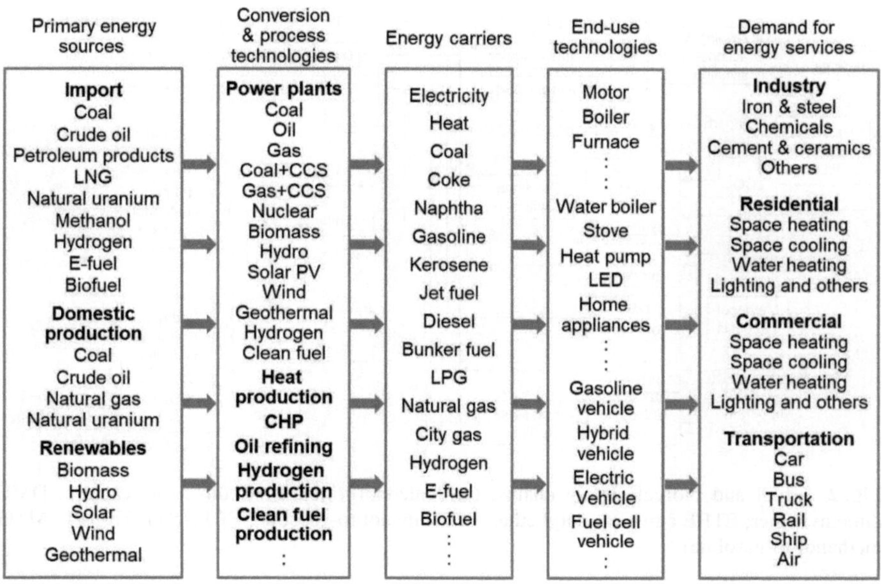

Fig. 1 Reference energy system

between 2010 and 2050 and simulate the energy transitions during this period. Figure 1 illustrates the model's reference energy system. The energy value chain in the MARKAL model consists of energy technologies (conversion, process, and end-use), energy sources and carriers (primary and secondary), and energy service demands. This version of the model used in this study incorporates into over 370 types of energy technologies, 120 kinds of energy sources and carriers, and 20 kinds of energy service demands.

Figure 2 shows the e-fuel and biofuel supply chains. This study assumes that the clean liquid fuels can be used as alternative fuels for transportation. E-fuels are imported from overseas or produced using CO_2 recovered from power plants or directly from the atmosphere via Fischer–Tropsch synthesis, the methanol-to-gasoline process, or dimethyl ether synthesis. E-fuels are converted into petroleum products such as gasoline, diesel fuel, and jet fuel via fractional distillation. Bioethanol is produced from biomass or imported. Two types of bioethanol production are considered; first-generation bioethanol, produced from edible energy crops (e.g., sugar, starch, and vegetable oil), and second-generation bioethanol, produced from inedible resources (e.g., cellulose, hemicellulose, and lignin). Bioethanol is converted into gasoline and jet fuel via ethyl tert-butyl ether synthesis and alcohol-to-jet processes, respectively. Electricity inputs to supply chain are assumed to be supplied by renewable power.

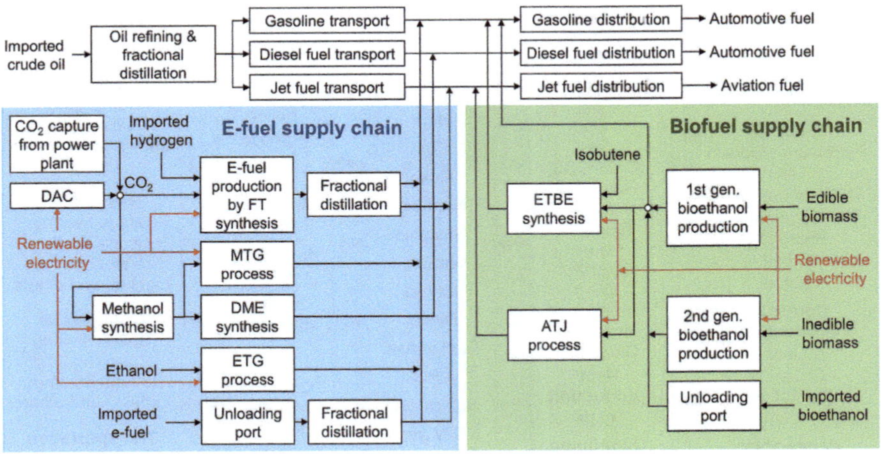

Fig. 2 E-fuel and biofuel supply chains. (ATJ alcohol-to-jet; DAC direct air capture; DME dimethyl ether; ETBE ethyl tert-butyl ether; ETG ethanol-to-gasoline; FT Fischer–Tropsch; MTG methanol-to-gasoline)

3 Results and Discussion

The transition of Japan's energy system with and without clean liquid fuels up to 2050 was simulated using the AIST-MARKAL model. In both cases, constraints on energy-derived CO_2 emissions followed Japan's long-term environmental goals. The future energy service demand followed the assumptions of the Institute of Energy Economics, Japan and the New Energy and Industrial Technology Development Organization [12].

Figure 3 displays the primary energy supply in the two cases. With clean liquid fuels, the total energy supply decreased from 20.2 EJ in 2010 to 18.1 EJ in 2050. Coal and oil imports were shown to decrease by 77% and 90%, respectively, over 40 years, while the renewable energy supplies show a 4.2-fold increase over the same period. Clean liquid fuel imports will increase moderately after 2025, reaching 1.8 EJ by 2050 and accounting for 10% of the total energy supply. Without clean liquid fuels, the supply of oil was observed to increase; oil imports in 2050 with and without clean liquid fuels are 0.9 EJ and 2.0 EJ, respectively.

Figure 4 displays power generation by source and the powertrain mix of passenger cars in the case of clean liquid fuels. The annual electricity supply will increase from 1104 TWh in 2010 to 1239 TWh in 2050 because the electrification of end-use technologies will outpace improvements in energy efficiency. In 2050, all electricity will be supplied by low-carbon sources: renewables, nuclear power, thermal power with CO_2 capture, or hydrogen. Among the power sources, renewables contribute the most to net zero emissions from power generation, accounting for 45% of the total amount of electricity generated in 2050. The powertrain mix has shown a rapid shift from conventional gasoline-fueled vehicles to electrified vehicles such

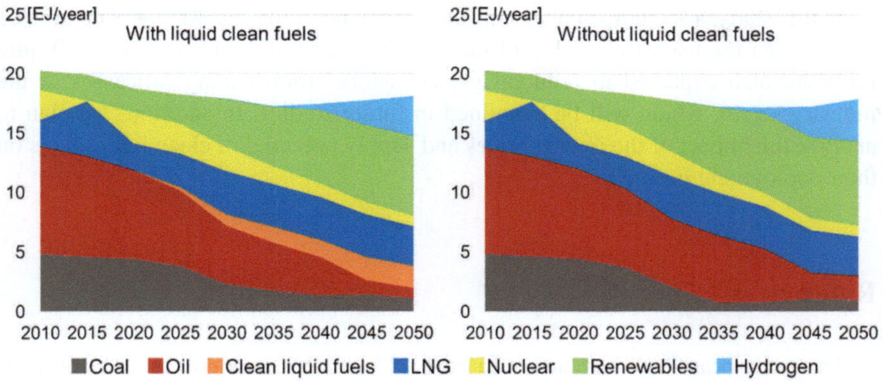

Fig. 3 Primary energy supply in the two cases

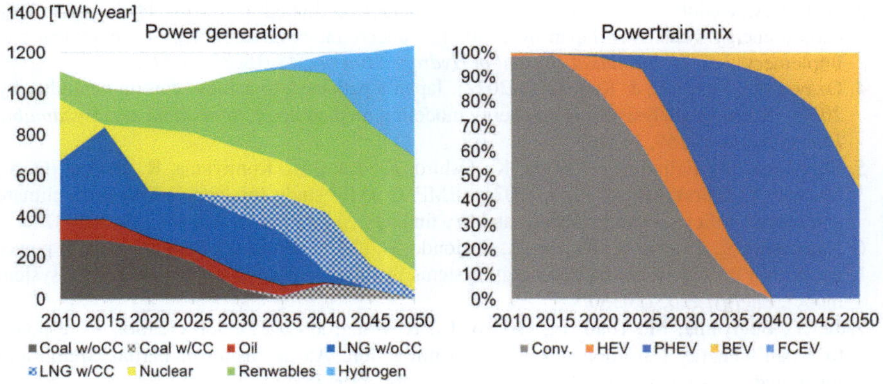

Fig. 4 Power generation by sources and powertrain mix of passenger cars in the case of clean liquid fuels. (CC CO_2 capture; Conv. conventional vehicles; HEV hybrid electric vehicle; PHEV plug-in hybrid electric vehicle; EV electric vehicle; FCEV fuel cell electric vehicle)

as hybrids (HEV), plug-in hybrids (PHEV), and electric vehicles (EV). By 2050, PHEVs and EVs will be fueled with low-carbon electricity and clean liquid fuels to reduce CO_2 emissions from passenger cars.

4 Conclusion

This study analyzed Japan's energy transition toward net-zero CO_2 emissions by 2050. The simulation results showed that, in 2050, clean liquid fuels will account for 10% of the primary energy supply and will be consumed as automotive fuels, which implies that clean liquid fuels will contribute to mitigating CO_2 emissions from the transportation sector.

In this study, a scenario analysis toward carbon neutrality by 2050 was conducted focusing on the use of clean liquid fuels in the transportation sector. Clean liquid fuels are also expected to reduce CO_2 emissions from heating processes in the industry sector, which will be examined in future studies. In addition, we plan to analyze the impact of the import prices and supply potential of clean liquid fuels on their deployment in Japan.

References

1. Intergovernmental Panel on Climate Change (IPCC). (2022). *Climate change 2022: Mitigation of climate change. Contribution of Working Group III to the sixth assessment report of the intergovernmental panel on climate change*. Cambridge University Press.
2. Agency for Natural Resources and Energy. (2021). *Strategic energy plan.*
3. Ozawa, A., Kudoh, Y., Murata, A., Honda, T., Saita, I., & Takagi, H. (2018). Hydrogen in low-carbon energy systems in Japan by 2050: The uncertainties of technology development and implementation. *International Journal of Hydrogen Energy, 43*, 18083–18094.
4. Ozawa, A., Tsani, T., & Kudoh, Y. (2022). Japan's pathways to achieve carbon neutrality by 2050 – Scenario analysis using an energy modeling methodology. *Renewable and Sustainable Energy Reviews, 169*, 112943.
5. Sugiyama, M., Fujimori, S., Wada, K., Oshiro, K., Kato, E., Komiyama, R., Herran, D. S., Matsuo, Y., Shiraki, H., & Ju, Y. (2021). EMF 35 JMIP study for Japan's long-term climate and energy policy: Scenario designs and key findings. *Sustainability Science, 16*, 355–374.
6. Nagatomo, Y., Ozawa, A., Kudoh, Y., & Hondo, H. (2021). Impacts of employment in power generation on renewable-based energy systems in Japan – Analysis using an energy system model. *Energy, 226*, 120350.
7. Ju, Y., Sugiyama, M., Kato, E., Oshiro, K., & Wang, J. (2022). Job creation in response to Japan's energy transition towards deep mitigation: An extension of partial equilibrium integrated assessment models. *Applied Energy, 318*, 119178.
8. Tokimatsu, K., Höök, M., McLellan, B., Wachtmeister, H., Murakami, S., Yasuoka, R., & Nishio, M. (2018). Energy modeling approach to the global energy-mineral nexus: Exploring metal requirements and the well-below 2 °C target with 100 percent renewable energy. *Applied Energy, 225*, 1158–1175.
9. Ozawa, A., Morimoto, S., Hatayama, H., & Anzai, Y. (2023). Energy–materials nexus of electrified vehicle penetration in Japan: A study on energy transition and cobalt flow. *Energy, 277*, 127698.
10. Energy technology systems analysis program homepage, https://iea-etsap.org/index.php/etsap-tools/model-generators/markal. Last accessed 2023/8/8.
11. Loulou, R., Goldstein, G., Noble, K. (2004). *Documentation for the MARKAL family of models.*
12. New Energy and Industrial Technology Development Organization. (2017). *Advancement of hydrogen technologies and utilization project analysis on comprehensive renewable energy systems (FY2014–FY2015) FY2015 annual report*. Tokyo, Japan.

Assessing Carbon Footprint Estimations of ChatGPT

Ithier d'Aramon, Boris Ruf, and Marcin Detyniecki

1 Introduction

ChatGPT, created by OpenAI, is a publicly accessible tool that utilizes the large language model GPT-3 [1]. ChatGPT's remarkable capability to produce language resembling that of humans and accomplish intricate tasks represents a noteworthy breakthrough within the realm of natural language processing and artificial intelligence. It is estimated to have reached 100 million monthly active users in January 2023, just 2 months after launch, making it the fastest-growing consumer application in history [2].

One drawback, however, is that training extensive neural networks like GPT-3 is known to entail considerable computational expenses, resulting in a significant demand for energy [3]. Recent accumulation of natural disasters has accelerated awareness of the unfolding climate crisis and increased the drive of policymakers and society to act [4, 5]. While the important role of energy-related greenhouse gases (GHGs) in climate mitigation is well understood, quantifying the specific environmental impact of goods and services remains complex [6]. Measuring the "carbon footprint" involves collecting data on GHGs from various sources, calculating their carbon dioxide equivalent and aggregating the results.

Work done during internship at AXA

I. d'Aramon
Work done during internship at AXA, Albert—Business & Data School, Paris, France
e-mail: idaramon@albertschool.com

B. Ruf (✉) · M. Detyniecki
AXA Group Operations, AI Research, Paris, France
e-mail: boris.ruf@axa.com; marcin.detyniecki@axa.com

P. Pong (ed.), *Renewable Energy Resources and Conservation*, Green Energy and Technology, https://doi.org/10.1007/978-3-031-59005-4_15

In particular the incoming Artificial Intelligence Act in the European Union could give the matter a boost. In Article 28b 2. (d) of an amendment which was adopted in June 2023 by the European Parliament, providers of large language models (foundation models) are explicitly required to "mak[e] use of applicable standards to reduce energy use, resource use and waste, as well as to increase energy efficiency, and the overall efficiency of the system. [...] They shall be designed with capabilities enabling the measurement and logging of the consumption of energy and resources, and, where technically feasible, other environmental impact the deployment and use of the systems may have over their entire lifecycle" [7].

OpenAI has not disclosed any specific information regarding the energy consumption or carbon emissions associated with its services. Therefore, various researchers have proposed estimations of their carbon footprint in order to narrow the information gap and support number-driven decision-making.

In the absence of official carbon emission reports by OpenAI, the carbon footprint of ChatGPT can only be estimated and extrapolated based on better studied systems with comparable features. However, carbon footprint modeling is a complex task that depends on many factors and includes several unknown variables. The proposed approximations take different approaches and their results deviate considerably. The presence of distinct underlying assumptions and variations in how the calculations are presented make it hard to evaluate and compare the propositions.

In this chapter, we study and compare three widely considered estimations of ChatGPT. In the next section, we briefly introduce our methodology. Afterward, we outline the approaches in detail. Finally, we compare the outcome and discuss the different estimates.

2 Methodology

We selected three carbon footprint studies for ChatGPT and reproduce each of them. For the in-depth analysis we use an open data model for carbon footprint quantification [8]. The advantage of formalizing the studies' calculations with such a methodology is to increase their accessibility. Using an open-source data viewer[1], which interprets the data records, the carbon footprint scenarios can be interactively explored in a web browser. The viewer also automatically handles unit conversion and finds a common ground based on the availability of emission data by type of emission. Another benefit of the data model is that references to the source of information for all underlying assumptions upgrade the documentation of the estimate and thus improve transparency and traceability.

It is important to note that we only examine the inference costs, i.e., the impact on the environment that occurs during the operation of the model. The also very

[1] https://github.com/borisruf/carbon-footprint-modeling-tool.

high energy consumption that occurs during the training process of the model is not considered in this chapter.

3 Carbon Scenarios

In this section, we replicate the results of three popular carbon footprint estimates for GPT-3, the language model used by ChatGPT. For better comparability, we consider for all scenarios the period of January 2023, when ChatGPT is reported to have had about 100M unique users who made 590M queries [9]. All data records created in the course of this study have been published on GitHub and are linked at the appropriate place in the text.

3.1 Scenario 1

Raghavenda Selvan, Assistant Professor at the University of Copenhagen, has estimated the carbon footprint of ChatGPT for Süddeutsche Zeitung, a German daily [10]. He concludes that emissions associated with the service could have accounted for 24.24 kt CO2e in January 2023 (see interactive online scenario with reproduced data model for details.[2])

In the absence of reliable data on the energy consumption of GPT-3, Selvan used the freely available open-source language model GPT-J as basis for his calculation [11]. To estimate the carbon footprint, Selvan ran experiments on a local workstation. The model was initialized in half-precision (16-bit float instead of 32-bit float) to fit into GPU memory. The average length of text generated by the model was 230 words. The energy consumption got measured with CarbonTracker, a tool developed by researchers from the University of Copenhagen [12]. As a result, one request to GPT-J consumed in average 0.01292 kWh.

Scaling to GPT-3, Selvan took the following factors into account: First, GPT-J only has 6 billion parameters, whereas GPT-3 has a much larger parameter count of 175 billion. Thus, Selvan assumes by a conservative estimate that the larger model could require 10x additional GPUs. Second, to fit the test environment, the GPT-J model was initialized in half-precision. The overhead of running a full-precision model compared to half-precision is considered about 1.5x more. In conclusion, Selvan estimates that submitting a query to GPT-3 requires 15x more energy compared to submitting a query to GPT-J.

Finally, Selvan uses the grid emission factor for Denmark in 2020 (0.212 kgC02e/kWh) to compute the GHGs.

[2] https://borisruf.github.io/carbon-footprint-modeling-tool/index.html?id=gpt-selvan-0.

3.2 Scenario 2

Chris Pointon also shared some consideration about the carbon footprint of ChatGPT [13]. Based on his calculation, the emissions linked to the service in January 2023 would amount to 225.64t CO2e (see interactive online scenario[3]).

Pointon bases his calculation on a statement made by Tom Goldstein, a Professor at the University of Maryland [14]. Based on experiments Goldstein had conducted with a modified BERT model, he estimates that if GPT-3 were executed on a single Nvidia A100 GPU, it would require approximately 350 milliseconds to generate a single word [15]. Accordingly, Pointon takes the documented energy consumption of this processor under full load (407 Watts per hour) and concludes that one request to generate a single word consumes 0.03957 Watts [16]. Finally, he assumes that each query to GPT-3 produces in average 30 words.

Pointon expects ChatGPT to be hosted in an Azure datacenter in California and therefore uses the grid emission factor for Western USA in 2021 (0.322167 kgC02e/kWh) to compute the emissions.

3.3 Scenario 3

Kasper Ludvigsen approximated the carbon footprint of ChatGPT in a blog post on Towards Data Science [17]. Based on his estimation, the emissions caused by ChatGPT could have accounted for 752.84 t CO2e in January 2023 (see interactive online scenario[4]).

Ludvigsen bases his calculation on a study by Alexandra Luccioni and others who investigated the carbon footprint of the BLOOM model [18]. With 176 billion parameters, this large language model is of similar size as GPT-3. The researchers measured the energy demand, while running the model during 18 days on 16 Nvidia A100 GPUs. In total, their system consumed 914 kWh, while handling 230,768 requests. Based on this information, Ludvigsen derives the average energy consumption for one query and extrapolates the total demand.

Similar to Pointon, Ludvigsen uses the grid emission factor for Western USA in 2021 (0.322167 kgC02e/kWh) to finally estimate the emissions.

[3] https://borisruf.github.io/carbon-footprint-modeling-tool/index.html?id=gpt-pointon-0.

[4] https://borisruf.github.io/carbon-footprint-modeling-tool/index.html?id=gpt-ludvigsen-0.

4 Comparison

A summary of key assumptions and results of the three scenarios can be found in Table 1. For a more interactive comparison, we refer to our online benchmark of the different approaches.[5]

Overall, it can be observed that the estimates are very far apart. Scenario 1 is the most pessimistic, estimating a carbon footprint of 24.24 kt CO2e. which is over 32 times higher than Scenario 3, which predicts 752.84 t CO2e, and even 107 times higher than Scenario 2 with 225.64 t CO2e.

We noticed that Selvan uses the grid emission factor for Denmark, while OpenAI's servers are more likely to be located in the USA, as also assumed by the other two authors. However, since the Danish electricity grid is actually less carbon-intensive, adjusting for this would only widen the gap.

On the other hand, Selvan assumes a response to include 230 words in average, while Pointon only calculates with 30 words. Adjusting the data model accordingly, Scenario 2 predicts emissions of 1.73 kt CO2e. Ludvigsen does not specify the number of words per response.

As basis for their estimations, all authors take the measured energy consumption of openly accessible large language models. Only Selvan conducted his own experiments, while the other two authors rely on results published by others. The model used by Selvan has fewer parameters than GPT-3, and he simply extrapolates his measurements by multiplying them by a factor. Pointon reconstructs the energy consumption via an estimated query duration by another researcher who extrapolated this value from a 3-billion parameter model. Ludvigsen takes a model of similar size as GPT-3 as reference.

All estimates calculate the average values of individual requests and then reuse these values to extrapolate the energy demand over a longer period of time. It is worth to note that such a practice is prone to accumulating rounding errors.

Overall, it is obvious that due to the many unknowns in the equation, the range of possible outcomes is very wide. Therefore, more details about the architecture behind ChatGPT as well as usage reports are urgently needed for making more accurate predictions.

Table 1 Key assumptions and results of the different scenarios

	Reference model	Words	Electricity grid	Energy per query (kWh)	Estimated emission in 01-2023 (CO2e)
1	GPT-J	230	Denmark (2020)	0.01292	24.24 kt
2	BERT variation	30	Western USA (2021)	0.00119	225.64 t
3	BLOOM	–	Western USA (2021)	0.00396	752.84 t

[5] https://borisruf.github.io/carbon-footprint-modeling-tool/benchmark.html?ids[]=gpt-selvan-0&ids[]=gpt-pointon-0&ids[]=gpt-ludvigsen-0.

5 Conclusion

We utilized an open data model for carbon footprint quantification to reproduce three carbon footprint estimates for ChatGPT. We show that this approach allows for a more efficient comparison of different estimation scenarios and also promotes discussions regarding their potential drawbacks. This finding emphasizes the importance of transparent and robust carbon footprint modeling.

References

1. Radford, A., & Narasimhan, K. (2018). *Improving language understanding by generative pre-training*. https://api.semanticscholar.org/CorpusID:49313245.
2. UBS. Has the AI rally gone too far?. Accessed on 21-06-2023. https://www.ubs.com/global/en/wealth-management/insights/chief-investment-office/house-view/daily/2023/latest-25052023.html.
3. Patterson, D., Gonzalez, J., Le, Q., Liang, C., Munguia, L. M., Rothchild, D., et al. (2021). *Carbon emissions and large neural network training*.
4. Stollberg, J., & Jonas, E. (2021). Existential threat as a challenge for individual and collective engagement: Climate change and the motivation to act. *Current Opinion in Psychology, 42*, 145–150. Psychology of Climate Change.
5. Masson, T., & Fritsche, I. (2021). We need climate change mitigation and climate change mitigation needs the 'We': A state-of-the-art review of social identity effects motivating climate change action. *Current Opinion in Behavioral Sciences, 42*, 89–96. Human Response to Climate Change: From Neurons to Collective Action
6. IEA. CO2 Emissions in 2022. Accessed on 21-06-2023. https://www.iea.org/reports/co2-emissions-in-2022.
7. European Parliament. Report on the proposal for a regulation of the European Parliament and of the Council on laying down harmonised rules on Artificial Intelligence (Artificial Intelligence Act) and amending certain Union Legislative Acts. Accessed on 21-06-2023. https://www.europarl.europa.eu/resources/library/media/20230516RES90302/20230516RES90302.pdf.
8. Ruf, B., & Detyniecki, M. (2022). Open and linked data model for carbon footprint scenarios. In *International conference on renewable energy and conservation (ICREC)*.
9. Milmo, D. ChatGPT reaches 100 million users two months after launch. The Guardian. Accessed on 22-06-2023. https://www.theguardian.com/technology/2023/feb/02/chatgpt-100-million-users-open-ai-fastest-growing-app.
10. Landwehr, T. Der Energiehunger der KIs. Süddeutsche Zeitung. Accessed on 21-06-2023. https://www.sueddeutsche.de/wissen/chat-gpt-energieverbrauch-ki-1.5780744.
11. Wang, B., & Komatsuzaki, A. (2021). *GPT-J-6B: A 6 billion parameter autoregressive language model*. https://github.com/kingoflolz/mesh-transformer-jax.
12. Anthony, L. F. W, Kanding, B., & Selvan, R. (2020). Carbontracker: Tracking and predicting the carbon footprint of training deep learning models. In *ICML workshop on challenges in deploying and monitoring machine learning systems*.
13. Pointon, C. (2022). The carbon footprint of ChatGPT. Medium. https://medium.com/@chrispointon/the-carbon-footprint-of-chatgpt-e1bc14e4cc2a.
14. Goldstein, T. Twitter. Accessed on 22-06-2023. https://twitter.com/tomgoldsteincs/status/1600196995389366274.
15. Geiping, J., & Goldstein, T. (2022). *Cramming: Training a language model on a single GPU in one day*.

16. Cloud Carbon Footprint. Accessed on 22-06-2023. https://www.cloudcarbonfootprint.org/docs/methodology/#appendix-iii-gpus-and-minmax-watts.
17. Ludvigsen, K. G. A. (2023). The carbon footprint of (ChatGPT. Medium. https://towardsdatascience.com/chatgpts-electricity-consumption-7873483feac4.
18. Luccioni, A. S., Viguier, S., & Ligozat, A. L. (2022). Estimating the Carbon Footprint of BLOOM, a 176B Parameter Language Model.

Part III
Waste-to-Energy and Microbial Fuel Cell Technology

New Fuel Source: Lemon Waste in MFCs-SC for the Generation of Bioelectricity

Santiago M. Benites, Rojas-Flores Segundo, Nazario-Naveda Renny, Nélida Milly Otiniano, Daniel Delfín-Narciso, and Cecilia V. Romero

1 Introduction

Different organic wastes are used to create by-products used as materials for cosmetics, food, energy sources, ethanol production, composting, etc. [1, 2]; it is being intensively investigated by different research groups around the world [3]. 1.3 million tons of solid waste generated in 2015 have been reported, and an increase of 60% has been estimated for the year 2025, turning this significant increase into a great problem for society because countries and governments do not have adequate policies for adequate collection of such a large amount of waste [4, 5]. Agricultural waste is the most influential in waste generation because specific environmental conditions are needed for its conservation, and often the farmers or businessmen dedicated to these items do not have this [6]. One of the most important fruits, due to its daily use in the preparation of meals, restaurants, and food stalls, is lemon, of which current production is estimated to be higher than the 16 million tons reported in 2016 (12.2 of world citrus production) [7–9]. This fruit is also being used in

S. M. Benites · R.-F. Segundo (✉) · N.-N. Renny
Vicerrectorado de Investigación, Universidad Autónoma del Perú, Lima, Peru
e-mail: santiago.benites@autonoma.pe; renny.nazario@autonoma.pe

N. M. Otiniano
Instituto de Investigación en Ciencias y Tecnología de la Universidad Cesar Vallejo, Trujillo, Peru
e-mail: notiniano@ucv.edu.pe

D. Delfín-Narciso
Grupo de Investigación en Ciencias Aplicadas y Nuevas Tecnologías, Universidad Privada del Norte, Trujillo, Peru
e-mail: daniel.delfin@upn.edu.pe

C. V. Romero
Facultad de medicina, Universidad Nacional de Trujillo, Trujillo, Peru
e-mail: cromerog@unitru.edu.pe

P. Pong (ed.), *Renewable Energy Resources and Conservation*, Green Energy and Technology, https://doi.org/10.1007/978-3-031-59005-4_16

medicine due to the important natural compounds they contain (flavonoids, citric acids, ascorbic acid, and essential oils) [10].

On the other hand, since the 1960s, microbial fuel cell (MFC) technology has gained significant importance due to the adaptability of different types of waste to be used as fuels in this technology [11, 12]. MFCs are composed of two chambers (cathodic and anodic), where the conversion of chemical energy to electrical energy occurs through the substrate used, and this is transferred from one side to the other through an external circuit [13, 14]. In this sense, reports have been observed on the use of different organic wastes used in single-chamber microbial fuel cells MFCs-SC as fuel; for example, Yaqoob et al. (2022) used rambutan, langsat, and mango waste as fuel, managing to generate peaks of 290, 220, and 400 mV respectively, in single-chamber fuel cells where graphite rods were used as electrodes [15]. Likewise, it has also been reported that the use of Dragon Fruit waste as fuel in single-chamber microbial fuel cells with zinc and copper electrodes generated voltage and current peaks of 0.46 ± 0.03 V and 2.86 ± 0.07 mA [16]. Papaya waste has also been reported as fuel, managing to generate voltage peaks and electric currents of 0.955 V and 5.079 mA, where they used sucrose to boost the electrical values and worked at a pH (optimal) of 4.98 [17]. In the literature, higher electrical values have been found in MFCs where metallic electrodes were used due to their high electrical conductivity and magnetic inherent electron transport properties [18].

The main objective of the research is to observe the potential of generating bioelectricity from lemon waste using microbial fuel cells using zinc and copper as electrodes. For this, the electric current, pH, electrical conductivity, and voltage values will be monitored for 15 days. Likewise, the importance of internal resistance, current density, and power density of microbial fuel cells will be measured. This research has a more appropriate use for lemon waste that can be generated in the different supply centers, companies, or farmers dedicated to producing and selling this product.

2 Materials and Methods

2.1 Manufacturing of MFCs-SC

Three SC-MFCs were made, obtained from the SAIDKOCC company (SAIDKOCC-10091720, Fujian, *China*) and the electrodes used were copper (Cu, 35 cm^2 area) and zinc (Zn, 55 cm^2 area) for the anode and tasted, respectively. An external circuit of copper wire (5.5 mm thick) with a resistance of 100 Ω was used, and Nafion 117 (Merck) was used as a proton exchange membrane, see Fig. 1.

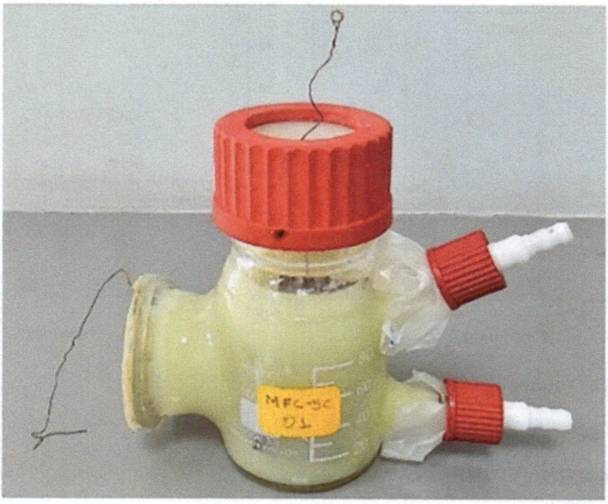

Fig. 1 MFCs-SC design

2.2 Collection of Lemon Waste

Lemon waste was collected from the Palermo -Ex Mayorista market, Trujillo, Peru; 1 kg of waste was selected for this. Subsequently, they were washed several times to eliminate any adhering contaminants from the environment in the laboratory. The waste was crushed (extractor, Labtron, LDO-B10-USA), obtaining approximately 600 mL of lemon waste extract.

2.3 Characterization of Microbial Fuel Cells

To measure the characterization of the electrical parameters (current and voltage), a multimeter (Prasek Premium PR-85, USA) with an external resistance (Rext.) of 100 Ω was used. For the PD (power density) and CD (current density), they were measured using the method described by Rojas-Flores et al. (2023) using as Rext. of 20 ± 0.3, 45 ± .8, 50 ± 4.9, 100 ± 10.2, 300 ± 18.5, 390 ± 22.5, 560 ± 30.5, 680 ± 60.5, 820 ± 7.5, 1000 ± 30 Ω [19]. Likewise, the pH and electrical conductivity values were monitored with a pH meter (110 Series Oakton, USA) and a conductivity meter (CD-4301, USA) during the 15 days of operation. Finally, the internal resistance of the MFCs-SC was found using the energy sensor (Vernier- ±30 V & ±1000 mA, USA) and Ohm's law. The MFCs connected in series managed to generate bioelectricity; this was achieved by connecting the cells with copper wire during the monitoring period.

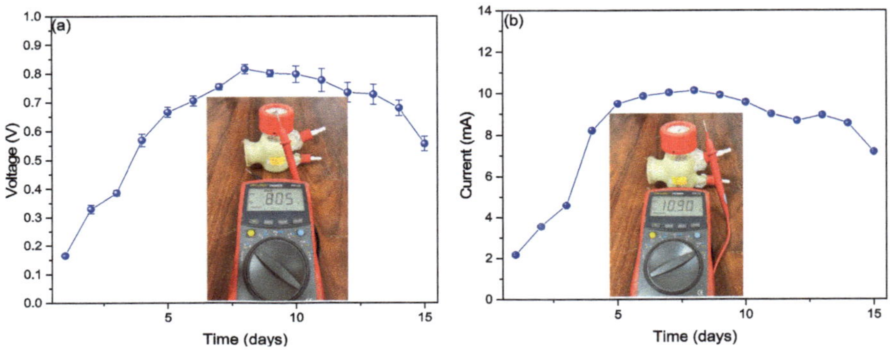

Fig. 2 Output (**a**) voltage and (**b**) current performance of SC-MFCs

3 Results and Analysis

The voltage values obtained from the MFCs-SC increased from the first day with a value of 0.164 ± 0.005 V until the eighth day (0.816 ± 0.017 V) after falling until the last day (0.556 ± 0.026 V) of monitoring, as it is observed in Fig. 2a. The voltage values obtained show a potential difference between the electrodes used due to the very characteristic of the substrate. This potential differential is due to the redox process that occurs in the initial moments of the experiment up to its maximum saturation point and then decays, due to which the compounds that originate these reactions begin to run out [20, 21]. Likewise, although the voltage values obtained in this research are relatively high compared to other studies, where they used a greater amount of substrate and cell sizes, future research must amplify the electrical values for better performance, that is, something similar to the research carried out by Koffi and Okale (2020), where they managed to amplify their values from 0.4 V to 99 ± 2 V [22]. The values of electric current have been reported in Fig. 2b, whose behavior is similar to the values obtained for voltage; that is, an increase is observed from day 1 (2.163 ± 0.004 mA) to day 8 (10.126 ± 0.093 mA) and then decayed until the last day (7.178 ± 0.101 mA). The current values increased from the first days due to the formation of the anodic biofilm and the transfer of electrodes from the anode to the cathode chamber through the external circuit [23]. In contrast, these values decreased because the environmental and nutrient conditions for microorganisms varied and decreased, respectively [24, 25].

The pH values shown by the MFCs-SC are observed in Fig. 3a, being able to appreciate small variations originated from day 1 of monitoring, whose increases are due to the same substrate processes that originated during power generation electric; It was possible to observe that the optimum operating pH (day 8) was 3 ± 0.12. Different pH values have been reported for different substrates and MFCs designs, each one different from the other because the microbial community present in the substrates varies depending on the environmental conditions and their nature [26, 27]. For example, Priya and Setty (2019) used apple juice waste as a substrate in

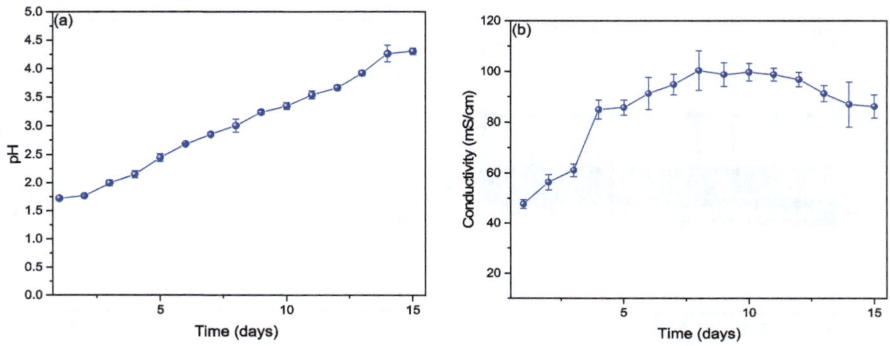

Fig. 3 Performance monitoring of (**a**) pH and (**b**) electrical conductivity values of SC-MFCs

SC-MFCs, which operated at a pH of 4.1, generating peak voltages of 0.40 V with an external resistance of 10,000 Ω [28]. So too, Iigatani et al. (2019), operating their MFCs-SC with shochu waste as a substrate at a pH of 4.5, managed to generate a power density of 0.44 W/m3 on the eighth day [29]. The values belonging to the electrical conductivity of the substrate are presented in Fig. 3b, showing an increase from the beginning (47,617 ± 1732 mS/cm) of the monitoring until the eighth day (100,362 ± 7810 mS/cm) to later show a decrease until the last day (86,101 ± 4582 mS/cm). The increase in values of electrical conductivity increases due to the reduction of the resistance of the MFCs-SC [30], this increase may be due in the first days to the high content of nutrients that fed the microorganisms, and when generating their metabolisms, they released electrons, which were captured by electrodes to generate electrical energy; similar events have been reported by other researchers [31–33]. At the same time, a decrease in the electrical conductivity values is due to the increase in the resistance of the MFCs-SC. It may be due to sedimentation in the final stages of monitoring [34].

Ohm's Law was used to calculate the internal resistance of the MFC-SC (Fig. 4a), where the voltage and current values were adjusted to the "*X*" and "*Y*" axes," respectively; Resulting in the slope of said adjustment, the internal resistance of the electronic device, the calculated internal resistance was 86.936 ± 14.505 Ω. The values of the internal resistance of an MFC have been attributed to the adhesion quality of the anodic biofilm on different types of substrates used in MFC; for example, Rincón et al. (2022) used banana debris as substrates in their MFCs-SC showing an internal resistance of 580.99 Ω, managing to generate a peak voltage and current of 286 mV and 0.2867 mA, respectively [35].

The internal resistance of MFCs-SC with cilantro debris has been reported, where they showed a Rint. of 75.581 ± 5.892 Ω, managing to generate voltage and current peaks of 0.882 ± 0.154 V and 2.287 ± 0.072 mA, respectively [36]. It has been observed that the MFCs where metallic electrodes were used to improve the electrical conductivity values of the system obtained lower internal resistance values, thus improving the performance of the MFCs [37, 38]. The power density (PD) values as a function of current density (CD) are observed in Fig. 4b, where the

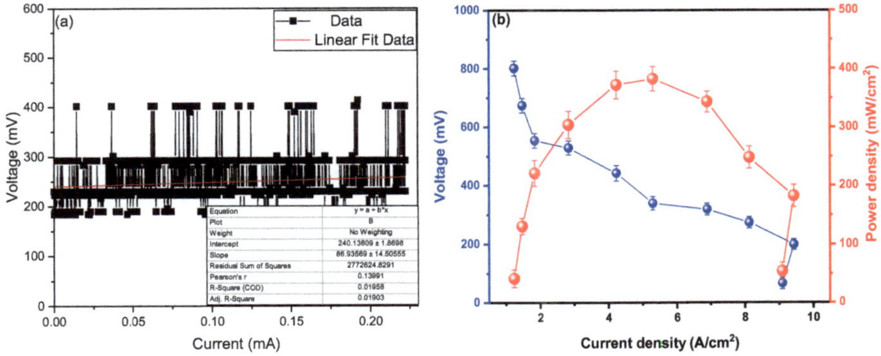

Fig. 4 Performance of the (**a**) internal resistance and (**b**) power density as a function of the current density of the SC-MFCs

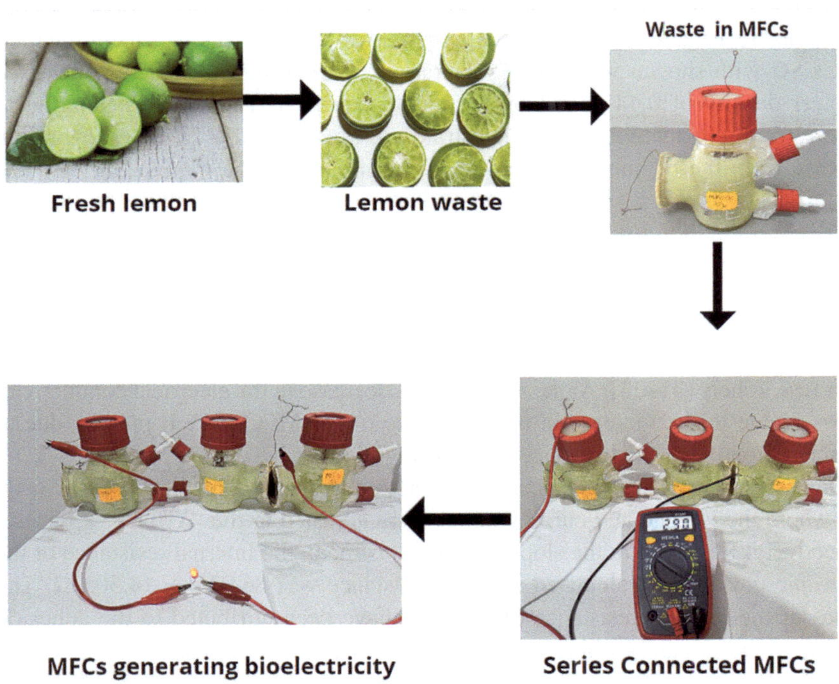

Fig. 5 Electric power generation process through lemon waste

maximum PD found was 384.365 ± 43.142 mW/cm^2 at a current density of 5.266 A/cm^2 with a maximum potential of 793.614 ± 19.461 mV. These values found are higher than those reported by Hussam et al. (2022), where I use bakery waste as substrates in MFCs and graphite electrodes, generating peaks of 1.001 mW/m^2 and 103.94 mA/m^2 of PD and CD, respectively [39]. Likewise, molasses waste has been used as a substrate in MFCs-SC with metallic electrodes, managing to generate PD and CD peaks of 5.45 ± 0.31 W/cm^2 and 308.06 mA/cm^2, respectively, attributing these values to the high conductivity of the electrodes used [40]. Figure 5 shows the process of generating electrical energy used, where the three MFCs-SC were connected in series, managing to create a voltage of 2.90 V, enough to turn on an LED (red).

4 Conclusions

The research was successfully completed, managing to generate the potential of lemon waste as fuel in MFCs-SC to produce bioelectricity. Likewise, high values of electric current and voltage were observed, with peak values of 0.816 ± 0.017 V and 10.126 ± 0.093 mA on the eighth day, in which they operated at an optimum pH of 3 ± 0.12, whose electrical conductivity of the substrate was 100.362. ± 7810 mS/cm. At the same time, the internal resistance shown was low (86.936 ± 14.505 Ω) compared to those in the literature, while the peak power density was 384.365 ± 43.142 mW/cm^2 at a current density of 5.266 A/cm^2 with a maximum output voltage of 793.614 ± 19.461 mV. Finally, the small bioelectricity generation process was schematized, where the MFCs-SC were placed in series, managing to generate 2.90 V, which was necessary to turn on an LED (red). For future work, it is recommended to cover the electrodes to avoid the oxidation observed in the metals, as well as to achieve greater cell power. On the other hand, the use and standardization of the optimal pH found in this research is recommended.

References

1. Shah, A. V., Singh, A., Mohanty, S. S., Srivastava, V. K., & Varjani, S. (2022). Organic solid waste: Biorefinery approach as a sustainable strategy in circular bioeconomy. *Bioresource Technology, 349*, 126835.
2. Siddiqui, S. A., Ristow, B., Rahayu, T., Putra, N. S., Yuwono, N. W., Mategeko, B., et al. (2022). Black soldier fly larvae (BSFL) and their affinity for organic waste processing. *Waste Management, 140*, 1–13.
3. Nguyen, M. K., Lin, C., Hoang, H. G., Sanderson, P., Dang, B. T., Bui, X. T., et al. (2022). Evaluate the role of biochar during the organic waste composting process: A critical review. *Chemosphere, 299*, 134488.
4. Karić, N., Maia, A. S., Teodorović, A., Atanasova, N., Langergraber, G., Crini, G., et al. (2022). Bio-waste valorisation: Agricultural wastes as biosorbents for removal of (in) organic pollutants in wastewater treatment. *Chemical Engineering Journal Advances, 9*, 100239.
5. Pandis, P. K., Kalogirou, C., Kanellou, E., Vaitsis, C., Savvidou, M. G., Sourkouni, G., et al. (2022). Key points of advanced oxidation processes (AOPs) for wastewater, organic pollutants and pharmaceutical waste treatment: A mini review. *ChemEngineering, 6*(1), 8.
6. Amrul, N. F., Kabir Ahmad, I., Ahmad Basri, N. E., Suja, F., Abdul Jalil, N. A., & Azman, N. A. (2022). A review of organic waste treatment using black soldier fly (Hermetia illucens). *Sustainability, 14*(8), 4565.

7. Klimek-Szczykutowicz, M., Szopa, A., & Ekiert, H. (2020). Citrus limon (Lemon) phenomenon—A review of the chemistry, pharmacological properties, applications in the modern pharmaceutical, food, and cosmetics industries, and biotechnological studies. *Plants, 9*(1), 119.
8. Leoni, V. (2020). Stars vs lemons. Survival analysis of peer-to peer marketplaces: The case of Airbnb. *Tourism Management, 79*, 104091.
9. AlZamily, J. Y., & Naser, S. S. A. (2020). *Lemon classification using deep learning.*
10. Mansur, J., & Felix, B. (2021). On lemons and lemonade: The effect of positive and negative career shocks on thriving. *Career Development International, 26*(4), 495–513.
11. Ramya, M., & Kumar, P. S. (2022). A review on recent advancements in bioenergy production using microbial fuel cells. *Chemosphere, 288*, 132512.
12. Obileke, K., Onyeaka, H., Meyer, E. L., & Nwokolo, N. (2021). Microbial fuel cells, a renewable energy technology for bio-electricity generation: A mini-review. *Electrochemistry Communications, 125*, 107003.
13. Hoang, A. T., Nižetić, S., Ng, K. H., Papadopoulos, A. M., Le, A. T., Kumar, S., & Hadiyanto, H. (2022). Microbial fuel cells for bioelectricity production from waste as sustainable prospect of future energy sector. *Chemosphere, 287*, 132285.
14. Jadhav, D. A., Mungray, A. K., Arkatkar, A., & Kumar, S. S. (2021). Recent advancement in scaling-up applications of microbial fuel cells: From reality to practicability. *Sustainable Energy Technologies and Assessments, 45*, 101226.
15. Yaqoob, A. A., Guerrero-Barajas, C., Ibrahim, M. N. M., Umar, K., & Yaakop, A. S. (2022). Local fruit wastes driven benthic microbial fuel cell: A sustainable approach to toxic metal removal and bioelectricity generation. *Environmental Science and Pollution Research, 29*(22), 32913–32928.
16. Segundo, R. F., Benites, S. M., De La Cruz-Noriega, M., Vives-Garnique, J., Otiniano, N. M., Rojas-Villacorta, W., et al. (2023). Impact of dragon fruit waste in microbial fuel cells to generate friendly electric energy. *Sustainability, 15*(9), 7316.
17. Rojas-Flores, S., De La Cruz-Noriega, M., Benites, S. M., Delfín-Narciso, D., Luis, A. S., Díaz, F., et al. (2022). Electric current generation by increasing sucrose in papaya waste in microbial fuel cells. *Molecules, 27*(16), 5198.
18. Yaqoob, A. A., Ibrahim, M. N. M., & Rodríguez-Couto, S. (2020). Development and modification of materials to build cost-effective anodes for microbial fuel cells (MFCs): An overview. *Biochemical Engineering Journal, 164*, 107779.
19. Rojas-Flores, S., De La Cruz-Noriega, M., Cabanillas-Chirinos, L., Benites, S. M., Nazario-Naveda, R., Delfín-Narciso, D., et al. (2023). Use of Kiwi waste as fuel in MFC and its potential for use as renewable energy. *Fermentation, 9*(5), 446.
20. Prasad, J., & Tripathi, R. K. (2021). Scale-up and control the voltage of sediment microbial fuel cell for charging a cell phone. *Biosensors and Bioelectronics, 172*, 112767.
21. Syed, Z., Sonu, K., & Sogani, M. (2022). Cattle manure management using microbial fuel cells for green energy generation. *Biofuels, Bioproducts and Biorefining, 16*(2), 460–470.
22. Koffi, N., & Okabe, S. (2020). High voltage generation from wastewater by microbial fuel cells equipped with a newly designed low voltage booster multiplier (LVBM). *Scientific Reports, 10*(1), 1–9.
23. Rojas-Flores, S., De La Cruz-Noriega, M., Nazario-Naveda, R., Benites, S. M., Delfín-Narciso, D., Rojas-Villacorta, W., & Romero, C. V. (2022). Bioelectricity through microbial fuel cells using avocado waste. *Energy Reports, 8*, 376–382.
24. Zhang, L., Fu, G., & Zhang, Z. (2019). Electricity generation and microbial community in long-running microbial fuel cell for high-salinity mustard tuber wastewater treatment. *Bioelectrochemistry, 126*, 20–28.
25. Wang, Y., Lin, Z., Su, X., Zhao, P., Zhou, J., He, Q., & Ai, H. (2019). Cost-effective domestic wastewater treatment and bioenergy recovery in an immobilized microalgal-based photoautotrophic microbial fuel cell (PMFC). *Chemical Engineering Journal, 372*, 956–965.

26. Vélez-Pérez, L. S., Ramirez-Nava, J., Hernández-Flores, G., Talavera-Mendoza, O., Escamilla-Alvarado, C., Poggi-Varaldo, H. M., et al. (2020). Industrial acid mine drainage and municipal wastewater co-treatment by dual-chamber microbial fuel cells. *International Journal of Hydrogen Energy, 45*(26), 13757–13766.
27. Algar, C. K., Howard, A., Ward, C., & Wanger, G. (2020). Sediment microbial fuel cells as a barrier to sulfide accumulation and their potential for sediment remediation beneath aquaculture pens. *Scientific Reports, 10*(1), 1–12.
28. Priya, A. D., & Setty, Y. P. (2019). Cashew apple juice as substrate for microbial fuel cell. *Fuel, 246*, 75–78.
29. Iigatani, R., Ito, T., Watanabe, F., Nagamine, M., Suzuki, Y., & Inoue, K. (2019). Electricity generation from sweet potato-shochu waste using microbial fuel cells. *Journal of Bioscience and Bioengineering, 128*(1), 56–63.
30. Xu, H., Du, Y., Chen, Y., Wen, Q., Lin, C., Zheng, J., & Qiu, Z. (2022). Electricity generation in simulated benthic microbial fuel cell with conductive polyaniline-polypyrole composite hydrogel anode. *Renewable Energy, 183*, 242–250.
31. Rojas-Flores, S., Nazario-Naveda, R., Benites, S. M., Gallozzo-Cardenas, M., Delfín-Narciso, D., & Díaz, F. (2022). Use of pineapple waste as fuel in microbial fuel cell for the generation of bioelectricity. *Molecules, 27*(21), 7389.
32. Moradian, J. M., Mi, J. L., Dai, X., Sun, G. F., Du, J., Ye, X. M., & Yong, Y. C. (2022). Yeast-induced formation of graphene hydrogels anode for efficient xylose-fueled microbial fuel cells. *Chemosphere, 291*, 132963.
33. Tahir, C. A., Pásztory, Z., Agarwal, C., & Csóka, L. (2022). Electricity generation and wastewater treatment with membrane-less microbial fuel cell. In *Application of microbes in environmental and microbial biotechnology* (pp. 235–261).
34. Thipraksa, J., & Chaijak, P. (2022). Improved the coconut shell biochar properties for bio-electricity generation of microbial fuel cells from synthetic wastewater. *Journal of Degraded & Mining Lands Management, 9*(4), 3613.
35. Rincón-Catalán, N. I., Cruz-Salomón, A., Sebastian, P. J., Pérez-Fabiel, S., Hernández-Cruz, M. D. C., Sánchez-Albores, R. M., et al. (2022). Banana waste-to-energy valorization by microbial fuel cell coupled with anaerobic digestion. *PRO, 10*(8), 1552.
36. Rojas-Flores, S., De La Cruz-Noriega, M., Cabanillas-Chirinos, L., Nazario-Naveda, R., Gallozzo-Cardenas, M., Diaz, F., & Murga-Torres, E. (2023). Potential use of coriander waste as fuel for the generation of electric power. *Sustainability, 15*(2), 896.
37. Zhang, G., Liang, D., Zhao, Z., Qi, J., & Huang, L. (2022). Enhanced performance of microbial fuel cell with electron mediators from tetracycline hydrochloride degradation. *Environmental Research, 206*, 112605.
38. Erensoy, A., Mulayim, S., Orhan, A., Cek, N., Tuna, A., & Ak, N. (2022). The system design of the peat-based microbial fuel cell as a new renewable energy source: The potential and limitations. *Alexandria Engineering Journal, 61*(11), 8743–8750.
39. Hussain, F., Al-Zaqri, N., Adnan, A. B. M., Hussin, M. H., Oh, S. E., & Umar, K. (2022). Impact of bakery waste as an organic substrate on microbial fuel cell performance. *Sustainable Energy Technologies and Assessments, 53*, 102713.
40. Flores, S. R., Pérez-Delgado, O., Naveda-Renny, N., Benites, S. M., De La Cruz-Noriega, M., & Narciso, D. A. D. (2022). Generation of bioelectricity using molasses as fuel in microbial fuel cells. *Environmental Research, Engineering and Management, 78*(2), 19–27.

Eco-friendly Generation of Electricity Using the *Bacteria Proteus Vulgaris* as a Catalyst

Santiago M. Benites, Rojas-Flores Segundo, De La Cruz-Noriega Magaly, Nazario-Naveda Renny, Nélida Milly Otiniano, and Daniel Delfín-Narciso

1 Introduction

The recent advances in the generation of electrical energy in an unconventional way have generated great achievements in the area of renewable energy [1]. Within the wide range of technologies for electricity generation are microbial fuel cells (MFCs), which are found in different designs, materials and types; but their common operation is based on two chambers (anodic and cathodic), each with their respective electrodes that are joined on the outside by an electrical circuit through which the electrons go from the anodic chamber to the cathodic. Furthermore, the chambers are separated by a proton exchange membrane [2–5]. Electricity is generated in several ways, one of the most important is through oxidation-reduction chemical reactions, which can also be produced by microorganisms [6, 7]. Microorganisms are responsible for generating the biofilm on the electrodes. A large number of these microorganisms responsible for releasing electrons in the process of oxidizing the organic matter present in the MFCs substrate have been identified [8–10]. MFCs offer several operational and functional advantages over the technologies currently used to generate energy from organic matter, among which we can mention: They

S. M. Benites · R.-F. Segundo (✉) · D. L. C.-N. Magaly · N.-N. Renny
Vicerrectorado de Investigación, Universidad Autónoma del Perú, Lima, Peru
e-mail: santiago.benites@autonoma.pe; mdelacruzn@autonoma.edu.pe; renny.nazario@autonoma.pe

N. M. Otiniano
Instituto de Investigación en Ciencias y Tecnología de la Universidad Cesar Vallejo, Trujillo, Peru
e-mail: notiniano@ucv.edu.pe

D. Delfín-Narciso
Grupo de Investigación en Ciencias Aplicadas y Nuevas Tecnologías, Universidad Privada del Norte, Trujillo, Peru
e-mail: daniel.delfin@upn.edu.pe

P. Pong (ed.), *Renewable Energy Resources and Conservation*, Green Energy and Technology, https://doi.org/10.1007/978-3-031-59005-4_17

provide high efficiency in the direct conversion of energy from the substrate into electricity. They work efficiently at room temperature and even at low temperatures. They do not require gas treatment, in this case, the exhaust gases are enriched in carbon dioxide. They do not need energy input for aeration. They have the potential for widespread application in places lacking electrical infrastructure and also to broaden the diversity of fuels available to meet energy needs [11, 12]. In this way, they contribute to solving both the energy crisis and environmental pollution [13]. One of its most useful and immediate applications is the treatment of wastewater for the production of renewable energy [14].

Broadly, microorganisms have an important role in the use of wastewater, organic compounds, or other types of waste as fuel in the different types of MFCs [15, 16]; they are mainly responsible for the generation of electrical current when they transform the chemical energy contained in organic matter into energy electrical [17–19]. Bacteria can use soluble components that physically transport the electron from an (intra)cellular compound, which is oxidized, to the electrode surface; they can also produce their own redox mediators, through the production of reversibly reducible organic compounds (secondary metabolites). or through the generation of oxidizable metabolites (primary metabolites) [11]. In an MFC, the electron crosses the cell membrane of the bacterium, where it is picked up by a mediator molecule and brought to the anode. In turn, these electrons produced in the oxidation process of organic and inorganic matter travel from the anode to the cathode, where they combine with protons and oxygen from the air to produce water molecules (reduction process). MFCs must use a separator (membranes or salt bridge) that allows proton transport from the anode compartment to the cathode and restricts oxygen transport to maintain anoxic conditions in the anode chamber [12]. The microorganisms that transfer electrons to a solid anode are also called exoelectrogens and microorganisms that accept electrons are called electrophytes [20–22].

Due to the increase in research in this area, new species that can generate electricity in MFCs have been reported; however, few are the strains that can generate large voltage potentials like those that generate mixed communities [23, 24]. Bacterial communities used as fuels in MFCs have been reported to observe their potential, for example, Hassan et al. (2019) used the *Brevibacillus borstelensis* bacteria as fuel, managing to generate 990 $\pm$ 5 mV voltage peaks and 188.5 mW/m^2 of power density, those values decreased until the end of the monitoring [25]. It has also been found that the *Saccharomyces cerevisiae* yeast has been used in single-chamber MFCs, reaching approximately 0.76 V and 1.43 mA of peak voltage and current, respectively, using zinc and copper electrodes, showing the importance of this type of metallic electrodes [26]. Likewise, *Paenibacillus lautus* has been used as a substrate in single-chamber MFCs with carbon felt electrodes, managing to generate power density peaks of 212.4 $\pm$ 29.2 mW/m^2 [27]. In this sense, Jamlus et al. (2021) used four different types of microorganisms (*Bacillus licheniformis, Bacillus velezensis, Klebsiella pneumonia,* and *K. variicola*) in single-chamber MFCs, observing that *K. pneumonia* is the one that generated higher voltage values (242 mV) [28].

In this sense, the objective of this research is to generate bioelectricity in microbial fuel cells with a single chamber, using the *Proteus vulgaris* bacteria as fuel (substrate) for a period of 30 days. For this, the values of voltage, current, and pH of the MFCs were monitored; thus, the internal resistance, power, current density, and power density were also calculated. Finally, the micrographs by SEM (scanning electron microscope) of the biofilm formed at the end of the monitoring were observed. In this way, it was possible to know the potential of this bacteria to generate electrical current, and to use the wastewater where this bacterium is found as fuel, to generate electrical energy in an environmentally friendly way.

2 Materials and Methods

2.1 *Fabrication of Microbial Fuel Cells*

250 mL polyvinyl chloride (PVC) containers were used as anodic and cathodic chambers, while for the cathodic electrode (copper, Cu) a circular hole of 10 cm in diameter was made on one of the faces of the container and the electrode anodic, a 12.5 cm^2 zinc plate was used in the center of the container, as shown in Fig. 1. A concentration of 6 g of KCl was used plus 14 g of agar and 400 mL of H_2O for a proton exchange membrane of 10 mL.

2.2 *Characterization of Microbial Fuel Cells*

The values of voltage and electrical current observed were measured with a multimeter (Prasek Premium PR-85) and those of pH by means of a pH-meter 110 Series Oakton, all this for a period of 30 days. While the power density (PD) and current density (CD) values were obtained based on what was done by Segundo et al. (2022) [29], while the resistance and power values of the MFCs were measured using an energy sensor (Vernier- ±30 V and ±1000 mA).

2.3 *Reactivation of the Bacterial Strain*

A pure culture of *P. vulgaris* was used. This bacterium was previously isolated and identified from tomato extract samples. The characteristics of its characterization in BLAST are shown in Table 1, where it can be seen that the sequences found have a percentage of identity of 100% with the genus Proteus.

From a pure culture, the bacterium was seeded in 10% sterile peptone water and incubated at 35 °C for 30 min. Then, it was planted in a sloping nutrient agar

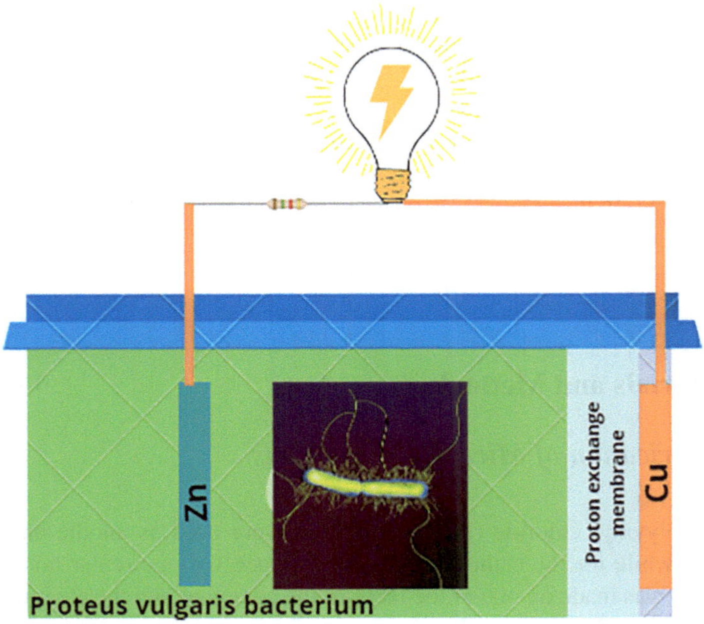

Fig. 1 Schematization of the MFC prototype

Table 1 Blast characterization of the rDNA sequence of the bacterium *P. vulgaris*

BLAST Characterization	Length of consensus sequence (nt)	% Maximum identified	Accession number	Phylogeny
Proteus vulgaris	1467	100.00	CP023965.1	Celullar organisms; Bacteria; Proteobacteria; Gammaproteobacteria; Enterobacterales; Morganellaceae; Proteus

medium and incubated at 35 °C for 24 h, in order to obtain pure cultures to prepare the inoculums.

2.4 *Preparation of the Inoculum of Proteus Vulgaris*

Pure cultures were seeded in 200 mL of nutrient broth and incubated at 35 °C for 24 h. Immediately, the biomass was separated by centrifugation at 4800 rpm for 10 min. The obtained precipitate was washed twice with sterile physiological saline solution (s.p.s.s) and finally resuspended in the same solution. This resuspension was used as inoculum. For the preparation of the working volumes, a suspension was made in 300 mL of sterile distilled water, comparing it with the Mc Farland tube number 3 (9 x 108 CFU/ml). Then, 1.0 liters of minimal mineral medium buffered with phosphate buffer were prepared according to the following composition (g/L): $Na_2HPO4.6$; KH_2PO_4, 3; NH_4Cl, 1; NaCl, 0.5; $MgSO_4$-$7H_2O$, 0.246; $CaCl_2$, 0.01 [30]. Finally, the inoculum was added to 10% (20 ml) of the total volume (200 ml) to investigate the electrogenic capacity of the bacterium, which was carried out in triplicate.

3 Results and Analysis

Figure 2a shows the voltage values monitored during the 30 days, observing an increase from the first day (0.587 $\pm$ 0.005 V) to the eleventh day (0.9545 $\pm$ 0.0399 V) and then gradually decreasing until the last day. (0.157 $\pm$ 0.023 V). The increase in voltage values is mainly due to the fact that in the first few days the bacteria proliferate due to the medium used, while the decrease may be due to the sedimentation of dead bacteria over time [31, 32]. The values obtained in this investigation are higher than those reported in other investigations, for example, Patel et al. (2021) managed to generate maximum voltage peaks of 700 mV using Gram-positive *Paenibacillus* bacteria as a substrate and carbon brush electrodes, where the bacteria were subjected to various media to observe their potential, the most optimal being the minimum salt media [33]. Oxidation reactions in the anodic chamber decrease over time due to decreased substrate compounds [34]. The values of electrical current observed in the monitoring are shown in Fig. 2b; it was possible to observe that the values increased until the eleventh day (0.368 $\pm$ 0.0556 mA) and then decreased to 1.656 $\pm$ 0.082 mA on the last day of monitoring. The formation of the biofilm on the electrodes is the cause of the increase in electric current values, because it allows a greater conductivity of the electrons from the anodic to the cathodic chamber; this flow of electrons shown in the first days of monitoring is diminished in the last days due to not enriching the medium, the microorganisms present in the solution do not have energy sources for their metabolism [35, 36]. The pH values increased slightly from the first day of monitoring, staying in the acidic region as shown in Fig. 2c, with its optimum operating pH being 5.884 $\pm$ 0.07 on the eleventh day. The influence of this parameter directly influences the generation of electric current caused by the activity of microorganisms, many of these microorganisms need adequate pH for their proliferation [37]. The pH in many substrates has managed to work in acidic media and obtain voltage values greater than 1 volt; for microbial fuel cells at laboratory scales are really high [38, 39].

Figure 3a shows the values of the internal resistance ($R_{int.}$) of the fuel cells monitored for 30 days, being able to observe that the $R_{int.}$ average of the MFCs was 51,113 $\pm$ 4375 Ω; this value is really low, which would give us an answer to the high values of voltage and electrical current observed in Fig. 1a, b because any

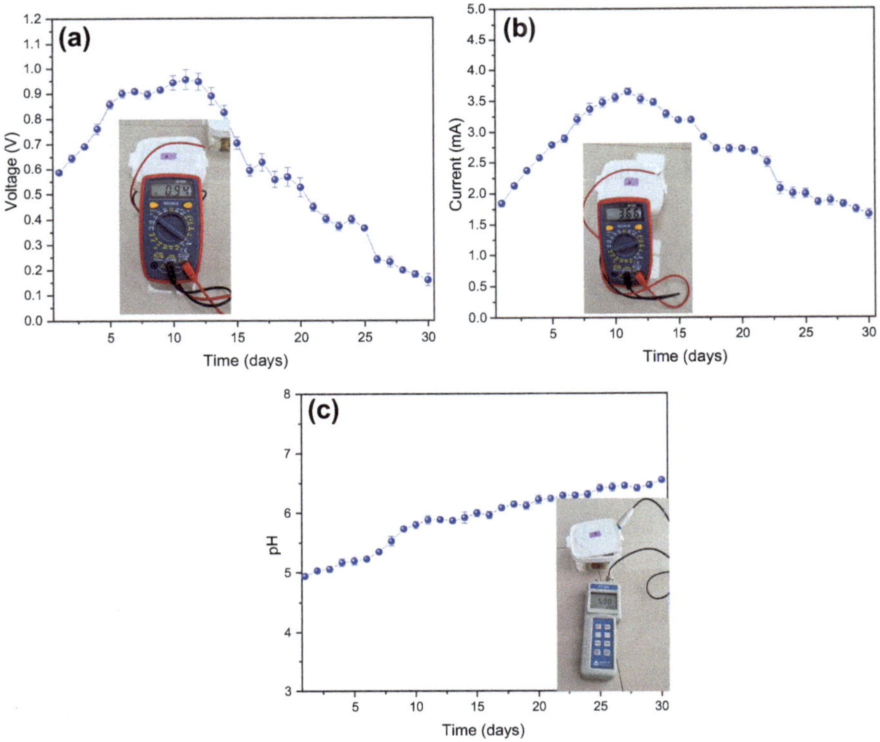

Fig. 2 Values of (**a**) voltage, (**b**) electric current, and (**c**) pH of microbial fuel cells

electrical device with low resistance values allows a better flow of electrons [40]. This is reinforced by the work done by Arkatkar et al. (2019) where in the MFCs studied they obtained different internal resistance values, generating voltage values directly proportional to their R_{int} [41]. In the same sense, Chen et al. (2021) managed to generate voltage peaks of 0.35 V in their MFCs with an internal resistance of 383.5 Ω, finding that the internal resistance increases in relation to the increase in the concentration of moxifloxacin hydrochloride used as a substrate in their research [42]. In Fig. 3b it is possible to observe the values of power density (PD) and voltage as a function of current density (CD), generating $\text{PD}_{\text{max.}}$ of 236.54 ± 6.324 mW/cm^2 in a DC of 5.469 A/cm^2 at a peak voltage of 889.23 ± 25.16 V. The values generated in this investigation exceed those carried out by other works, for example, Simeon et al. (2020) used human urine as substrates in their MFCs, managing to generate PD peaks of 85.76 mW/m^2 at a DC of 215.53 mA/m^2 with a peak voltage of approximately 780 mV on day 20 of operation [43]. Likewise, Christwardana et al. (2021) managed to observe that the power density values can increase if the O2 at the cathode increases, but up to a certain saturation point and that for each substrate used the saturation point is different [44]. Rajesh et al. reported PD values

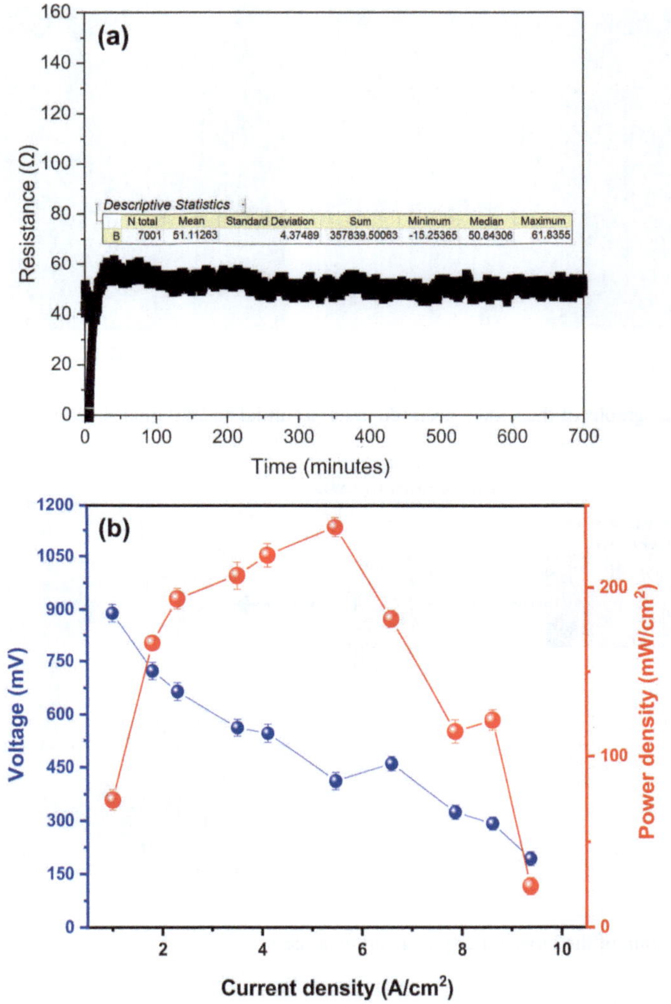

Fig. 3 Values of (**a**) internal resistance and (**b**) power density as a function of current density

of 21.62 W/m^2 at a DC of 66.35 mA/m^2 using Chaetoceros (15% concentration) as substrate in their MFCs, finding that there are optimal substrate concentrations [45].

Figure 4a shows the micrograph of the anodic electrode in its initial state, where a smooth surface is observed, while Fig. 4b shows the micrograph of the same electrode in its final state with an irregular surface and porous, this is possibly due to the adhesion of microorganisms in the creation of the biofilm. The same was observed in the research carried out by Saha et al. (2019) mentioning that the change of the surface is due to the fact that the bacterial community created a biofilm [46].

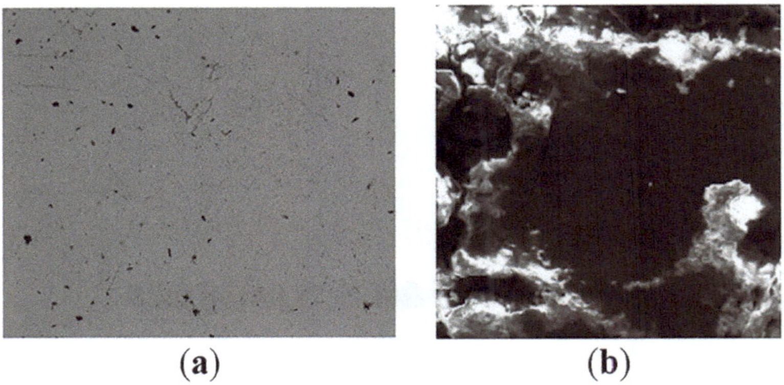

Fig. 4 Micrographs of the anode electrode in its (**a**) initial and (**b**) final state of monitoring

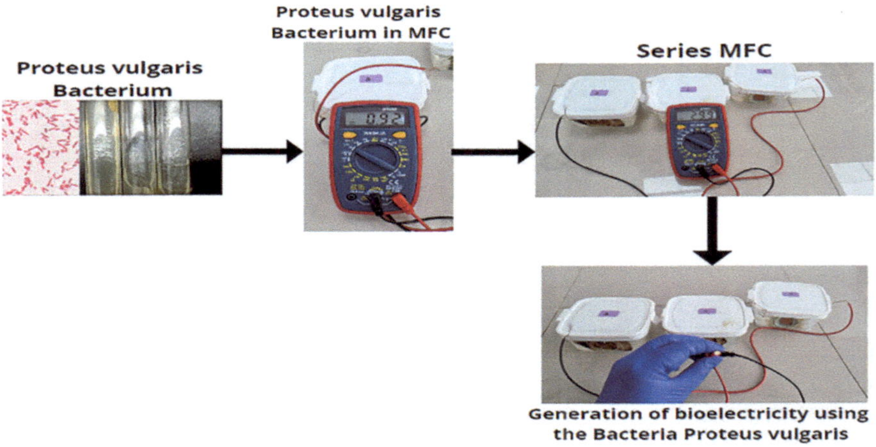

Fig. 5 Diagram of the bioelectricity generation process

Figure 5 shows the schematization of the bioelectricity generation process using the *P. vulgaris* bacteria as fuel, the microbial fuel cells connected in series are shown, which were capable of generating 2.99 V, enough to light a led spotlight (red).

4 Conclusions

This research managed to successfully generate bioelectricity, using *P. vulgaris* bacteria (molecularly identified) as fuel (substrate) in a single-chamber microbial fuel cells on a laboratory scale using zinc and copper electrodes, managing to generate maximum values of voltage and electric current of 0.9545 ± 0.0399 V and 0.368 ± 0.0556 mA on the eleventh day, operating with an optimum pH of

5.884 ± 0.07. The internal resistance found was 51.113 ± 4.375 Ω, considerably low, which is why the power density was 85.76 mW/m^2 in a current density of 215.53 mA/m^2 with a voltage of 780 mV. The micrographs show a smooth surface in its initial state, but at the end of the monitoring the formation of the biofilm is observed on the anode electrode. Finally, the microbial fuel cells were connected in series, managing to turn on a red LED during monitoring.

For future work, it is recommended to cover the metallic electrodes with some type of chemical compound that is not harmful to bacteria, or to work with other types of metallic electrodes to take advantage of their property as an excellent electrical conductor, as well as standardizing the pH value, to the optimal value to generate the maximum potential of microbial fuel cells in all their operationalization.

References

1. Sinsel, S. R., Riemke, R. L., & Hoffmann, V. H. (2020). Challenges and solution technologies for the integration of variable renewable energy sources—A review. *Renewable Energy, 145*, 2271–2285.
2. Palanisamy, G., Jung, H. Y., Sadhasivam, T., Kurkuri, M. D., Kim, S. C., & Roh, S. H. (2019). A comprehensive review on microbial fuel cell technologies: Processes, utilization, and advanced developments in electrodes and membranes. *Journal of Cleaner Production, 221*, 598–621.
3. Yaqoob, A. A., Ibrahim, M. N. M., & Guerrero-Barajas, C. (2021). Modern trend of anodes in microbial fuel cells (MFCs): An overview. *Environmental Technology & Innovation, 23*, 101579.
4. Mashkour, M., Rahimnejad, M., Raouf, F., & Navidjouy, N. (2021). A review on the application of nanomaterials in improving microbial fuel cells. *Biofuel Research Journal, 8*(2), 1400–1416.
5. Arun, S., Sinharoy, A., Pakshirajan, K., & Lens, P. N. (2020). Algae based microbial fuel cells for wastewater treatment and recovery of value-added products. *Renewable and Sustainable Energy Reviews, 132*, 110041.
6. Munoz-Cupa, C., Hu, Y., Xu, C., & Bassi, A. (2021). An overview of microbial fuel cell usage in wastewater treatment, resource recovery and energy production. *Science of the Total Environment, 754*, 142429.
7. Gupta, S., Srivastava, P., Patil, S. A., & Yadav, A. K. (2021). A comprehensive review on emerging constructed wetland coupled microbial fuel cell technology: Potential applications and challenges. *Bioresource Technology, 320*, 124376.
8. Kumar, S. S., Kumar, V., Malyan, S. K., Sharma, J., Mathimani, T., Maskarenj, M. S., et al. (2019). Microbial fuel cells (MFCs) for bioelectrochemical treatment of different wastewater streams. *Fuel, 254*, 115526.
9. Shabani, M., Younesi, H., Pontié, M., Rahimpour, A., Rahimnejad, M., & Zinatizadeh, A. A. (2020). A critical review on recent proton exchange membranes applied in microbial fuel cells for renewable energy recovery. *Journal of Cleaner Production, 264*, 121446.
10. Kabutey, F. T., Zhao, Q., Wei, L., Ding, J., Antwi, P., Quashie, F. K., & Wang, W. (2019). An overview of plant microbial fuel cells (PMFCs): Configurations and applications. *Renewable and Sustainable Energy Reviews, 110*, 402–414.
11. Rabaey, K., & Verstraete, W. (2005). Microbial fuel cells: Novel biotechnology for energy generation. *Trends in Biotechnology, 23*, 291–298. https://doi.org/10.1016/j.tibtech.2005.04.008

12. Mora Collazos, A., & Bravo Montaño, E. (2017). Diversidad Bacteriana Asociada a Biopelículas Anódicas En Celdas de Combustible Microbianas Alimentadas Con Aguas Residuales. *Acta Biolo. Colomb., 22*, 77. https://doi.org/10.15446/abc.v22n1.55766
13. Hoang, A. T., Nižetić, S., Ng, K. H., Papadopoulos, A. M., Le, A. T., Kumar, S., Hadiyanto, H., & Pham, V. V. (2022). Microbial fuel cells for bioelectricity production from waste as sustainable prospect of future energy sector. *Chemosphere, 287*, 132285. https://doi.org/10.1016/j.chemosphere.2021.132285
14. Logan, B. E., Hamelers, B., Rozendal, R., Schröder, U., Keller, J., Freguia, S., et al. (2006). Combustible Microbianas: Metodología y Tecnología. *Ciencia y tecnología ambiental, 40*, 5181–5192.
15. Enamala, M. K., Dixit, R., Tangellapally, A., Singh, M., Dinakarrao, S. M. P., Chavali, M., et al. (2020). Photosynthetic microorganisms (Algae) mediated bioelectricity generation in microbial fuel cell: Concise review. *Environmental Technology & Innovation, 19*, 100959.
16. Konovalova, E. Y., Stom, D. I., Zhdanova, G. O., Yuriev, D. A., Li, Y., Barbora, L., & Goswami, P. (2018, April). The microorganisms used for working in microbial fuel cells. In *AIP conference proceedings* (Vol. 1952, No. 1, p. 020017). AIP Publishing LLC.
17. Ulusoy, I., & Dimoglo, A. (2018). Electricity generation in microbial fuel cell systems with Thiobacillus ferrooxidans as the cathode microorganism. *International Journal of Hydrogen Energy, 43*(2), 1171–1178.
18. Mekuto, L., Olowolafe, A. V., Huberts, R., Dyantyi, N., Pandit, S., & Nomngongo, P. (2020). Microalgae as a biocathode and feedstock in anode chamber for a selfsustainable microbial fuel cell technology: A review. *South African Journal of Chemical Engineering, 31*(1), 7–16.
19. Almatouq, A., Babatunde, A. O., Khajah, M., Webster, G., & Alfodari, M. (2020). Microbial community structure of anode electrodes in microbial fuel cells and microbial electrolysis cells. *Journal of Water Process Engineering, 34*, 101140.
20. Stöckl, M., Teubner, N. C., Holtmann, D., Mangold, K. M., & Sand, W. (2019). Extracellular polymeric substances from Geobacter sulfurreducens biofilms in microbial fuel cells. *ACS Applied Materials & Interfaces, 11*(9), 8961–8968.
21. Ramya, M., & Kumar, P. S. (2022). A review on recent advancements in bioenergy production using microbial fuel cells. *Chemosphere, 288*, 132512.
22. Choudhury, P., Ray, R. N., Bandyopadhyay, T. K., Basak, B., Muthuraj, M., & Bhunia, B. (2021). Process engineering for stable power recovery from dairy wastewater using microbial fuel cell. *International Journal of Hydrogen Energy, 46*(4), 3171–3182.
23. Cao, Y., Mu, H., Liu, W., Zhang, R., Guo, J., Xian, M., & Liu, H. (2019). Electricigens in the anode of microbial fuel cells: Pure cultures versus mixed communities. *Microbial Cell Factories, 18*(1), 1–14.
24. Islam, M. A., Karim, A., Mishra, P., Dubowski, J. J., Yousuf, A., Sarmin, S., & Khan, M. M. R. (2020). Microbial synergistic interactions enhanced power generation in co-culture driven microbial fuel cell. *Science of the Total Environment, 738*, 140138.
25. Hassan, S. H., Abd el Nasser, A. Z., & Kassim, R. M. (2019). Electricity generation from sugarcane molasses using microbial fuel cell technologies. *Energy, 178*, 538–543.
26. Santiago, B., Rojas-Flores, S., De La Cruz Noriega, M., Cabanillas-Chirinos, L., Otiniano, N. M., Silva-Palacios, F., & Luis, A. S. (2020, July). Bioelectricity from Saccharomyces cerevisiae yeast through low-cost microbial fuel cells. In *Proceedings of the 18th LACCEI international multi-conference for engineering, education, and technology: Engineering, integration, and alliances for a sustainable development, virtual* (pp. 27–31).
27. Yu, Y. Y., Zhen, S. H., Chao, S. L., Wu, J., Cheng, L., Li, S. W., et al. (2022). Electrochemistry of newly isolated Gram-positive bacteria Paenibacillus lautus with starch as sole carbon source. *Electrochimica Acta, 411*, 140068.
28. Jamlus, N. I. I. M., Masri, M. N., Wee, S. K., & Shoparwe, N. F. (2021, May). Electricity generation by locally isolated electroactive bacteria in microbial fuel cell. In *IOP conference series: Earth and environmental science* (Vol. 765, No. 1, p. 012115). IOP Publishing.

29. Segundo, R. F., De La Cruz-Noriega, M., Milly Otiniano, N., Benites, S. M., Esparza, M., & Nazario-Naveda, R. (2022). Use of onion waste as fuel for the generation of bioelectricity. *Molecules, 27*(3), 625.
30. Kabir, M. M., Fakhruddin, A., Chowdhury, M., Pramanik, M. K., & Fardous, Z. (2018). Isolation and characterization of chromium (VI)-reducing bacteria from tannery effluents and solid wastes. *World Journal of Microbiology & Biotechnology, 34*(9), 126. https://doi.org/10.1007/s11274-018-2510-z
31. Rojas Flores, S. J., Benites, S. M., Agüero Quiñones, R., Enríquez-León, R., & Angelats Silva, L. (2020). Bioelectricity through microbial fuel cells from decomposed fruits using lead and copper electrodes. [Bioelectricidad mediante Celdas de Combustible Microbiana a partir de frutas descompuestas usando electrodos de plomo y cobre.].
32. Gallozzo Cardenas, M. M., Rojas-Flores, S., la Cruz-Noriega, D., Benites, S. M., Delfín-Narciso, D., Angelats-Silva, L., & Gallozzo Cárdenas, M. M. (2022). Electric current generation by increasing sucrose in papaya waste in microbial fuel cells. *Molecules, 27*(16), 5198.
33. Patel, D., Bapodra, S. L., Madamwar, D., & Desai, C. (2021). Electroactive bacterial community augmentation enhances the performance of a pilot scale constructed wetland microbial fuel cell for treatment of textile dye wastewater. *Bioresource Technology, 332*, 125088.
34. Cao, B., Zhao, Z., Peng, L., Shiu, H. Y., Ding, M., Song, F., et al. (2021). Silver nanoparticles boost charge-extraction efficiency in Shewanella microbial fuel cells. *Science, 373*(6561), 1336–1340.
35. Stokes, J. M., Lopatkin, A. J., Lobritz, M. A., & Collins, J. J. (2019). Bacterial metabolism and antibiotic efficacy. *Cell Metabolism, 30*(2), 251–259.
36. Wang, K., Mao, H., Wang, Z., & Tian, Y. (2018). Succession of organics metabolic function of bacterial community in swine manure composting. *Journal of Hazardous Materials, 360*, 471–480.
37. Leiva, E., Leiva-Aravena, E., Rodríguez, C., Serrano, J., & Vargas, I. (2018). Arsenic removal mediated by acidic pH neutralization and iron precipitation in microbial fuel cells. *Science of the Total Environment, 645*, 471–481.
38. Segundo, R. F., Magaly, D. L. C. N., Benites, S. M., Daniel, D. N., Angelats-Silva, L., Díaz, F., & Luis, C. C. (2022). Generation of electricity through papaya waste at different pH. *Environmental Research, Engineering & Management, 78*(4), 137.
39. De La Cruz-Noriega, M., Rojas-Flores, S., Nazario-Naveda, R., Benites, S. M., Delfín-Narciso, D., Rojas-Villacorta, W., & Diaz, F. (2022). Potential use of mango waste and microalgae Spirulina sp. for bioelectricity generation. *Environmental Research, Engineering and Management, 78*(3), 129–136.
40. Rossi, R., & Logan, B. E. (2020). Impact of external resistance acclimation on charge transfer and diffusion resistance in bench-scale microbial fuel cells. *Bioresource Technology, 318*, 123921.
41. Arkatkar, A., Mungray, A. K., & Sharma, P. (2019). Effect of microbial growth on internal resistances in MFC: A case study. In *Innovations in infrastructure* (pp. 469–479). Springer.
42. Chen, J., Wang, T., Zhang, K., Luo, H., Chen, W., Mo, Y., & Wei, Z. (2021). The fate of antibiotic resistance genes (ARGs) and mobile genetic elements (MGEs) from livestock wastewater (dominated by quinolone antibiotics) treated by microbial fuel cell (MFC). *Ecotoxicology and Environmental Safety, 218*, 112267.
43. Simeon, M. I., Asoiro, F. U., Aliyu, M., Raji, O. A., & Freitag, R. (2020). Polarization and power density trends of a soil-based microbial fuel cell treated with human urine. *International Journal of Energy Research, 44*(7), 5968–5976.
44. Christwardana, M., Yoshi, L. A., Setyonadi, I., & Maulana, M. R. (2021, April). Correlation between voltage, dissolved oxygen, and power density of yeast microbial fuel cell in different environmental waters as catholyte. In *AIP conference proceedings* (Vol. 2342, No. 1, p. 050001). AIP Publishing LLC.

45. Rajesh, P. P., Christine, P., & Ghangrekar, M. M. (2022). Optimum dose of Chaetoceros for controlling methanogenesis to improve power production of microbial fuel cell. *Water Science and Technology, 85*(1), 257–264.
46. Saha, T. C., Protity, A. T., Zohora, F. T., Shaha, M., Ahmed, I., Barua, E., et al. (2019). Microbial Fuel Cell (MFC) application for generation of electricity from dumping rubbish and identification of potential electrogenic bacteria. *Advances in Industrial Biotechnology, 2*, 10.

Multiple Block-Shaped Vertical Cathodes for Scale-Up of Floating Microbial Fuel Cells

Soichiro Hirose, Trang Nakamoto, and Kozo Taguchi

1 Introduction

Microbial fuel cells (MFC) are a renewable energy source that utilizes the biological activity of microbes, which release electrons when they decompose organic matter [1]. Since power is generated and organic matter is decomposed at the same time, MFC is an effective technology for creating a sustainable society [2]. However, MFC has yet to be put into practical use as much as other major renewable energies, such as solar and wind power [3].

A floating MFC (FMFC), studied recently, is an MFC in which an anode in liquid and an air cathode are combined into a single unit and floated in the liquid using a float [4–6]. By floating, a part of the cathode is let into the air, which improves the air cathode performance and prevents changes in the distance between the anode and the cathode due to changes in the water level [7]. In addition, FMFCs are expected to be suitable for practical use due to their ease of installation and maintenance.

Block-shaped electrodes have several advantages that make them more effective for practical applications than the commonly used sheet-shaped electrodes. For example, they have high physical stability and are suitable for mass production due to the simplicity of the fabrication method [8]. However, as the area of the block shape is increased, the possibility of cracking during the fabrication process increases, and physical stability is thought to decrease. Therefore, it is challenging to scale up FMFC by increasing the electrode area.

In this study, we developed an FMFC with multiple electrodes in order to scale up the FMFC with block-shaped electrodes. The electrode setup of a typical FMFC is horizontal to the liquid surface [9, 10]. However, the direction of the cathode

S. Hirose · T. Nakamoto · K. Taguchi (✉)
Ritsumeikan University, Kusatsu, Japan
e-mail: taguchi@se.ritsumei.ac.jp

P. Pong (ed.), *Renewable Energy Resources and Conservation*, Green Energy and Technology, https://doi.org/10.1007/978-3-031-59005-4_18

was made vertical to minimize the cut of the float area as much as possible. The relationship between the number of anodes and cathodes, respectively, and the output of the FMFC was investigated.

2 Materials and Methods

2.1 Block-Shaped Electrode Preparation

The block-shaped anodes and cathodes used in the experiment were fabricated in the same way as in the previous study [6], using rice husk smoked charcoal (Tokorozawa Ueki Bachi Center, Ltd., Saitama, Japan) and a Japanese ink called Bokuju (BA7-18, Kuretake Co., Ltd., Nara, Japan). The amount of rice husk smoked charcoal, and Bokuju used were 0.9 g and 3 ml for the anode and 0.6 g and 3 ml for the cathode, respectively. However, 0.1 g of copper was added to the cathode to prevent microbial attachment. The size of the electrode is $2 \times 2 \times 0.6$ cm.

2.2 FMFC Design with Varying Numbers of Electrodes

To investigate the relationship between the number of anodes and cathodes and FMFC output, four types of FMFCs were fabricated, each with a different number of anodes and cathodes, as shown in Fig. 1.

The fabricated FMFC was floated in a solution of 1 ml of muddy water collected from Japanese rice fields as a microbial source inoculated into an aqueous LB medium solution with a chemical oxygen demand of 2976 mg/l.

To further scale up, similar experiments were conducted by increasing the number of anodes and cathodes to four, as shown in Fig. 2a. We also developed an FMFC in which the anodes were also set up vertically to hold the anodes (Fig. 2b) more compactly. However, for practical use, 0.75 g of manganese oxide catalyst (Co-MnO_2/C) was added to the cathode instead of copper in this case.

2.3 Measurement of FMFC

FMFC was always connected by a 10 kΩ external resistor, except when measuring the power curve. The power curves were obtained by measuring the respective steady-state voltages when the external resistors connected were changed in the range of 10 kΩ to 100 Ω.

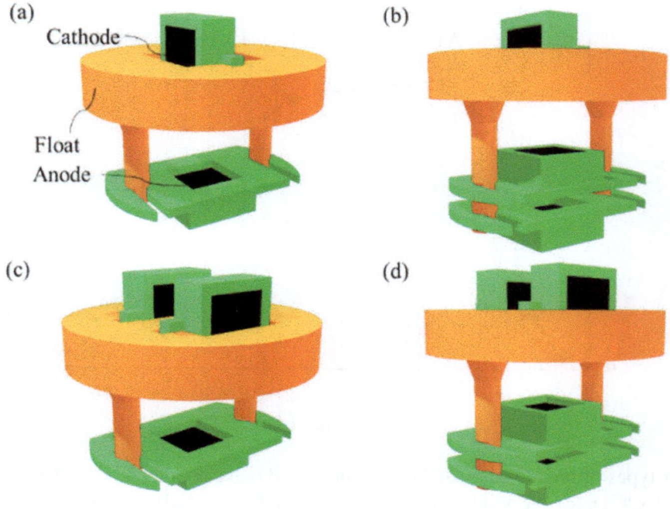

Fig. 1 Schematic of four fabricated FMFCs with (**a**) one anode and one cathode, (**b**) two anodes and one cathode, (**c**) one anode and two cathodes, and (**d**) two anodes and two cathodes

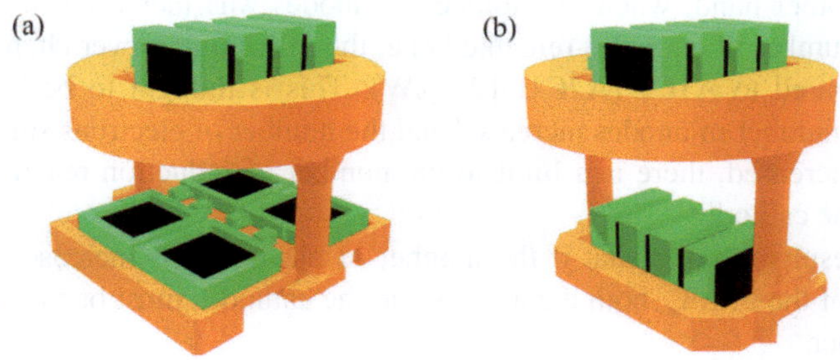

Fig. 2 Schematic of an FMFC with four electrodes where the anodes are set (**a**) horizontally or (**b**) vertically

3 Results and Discussion

3.1 Effect of the Number of Anodes and Cathodes on FMFC Output

Figure 3 shows that a conventional FMFC with a single anode and cathode (A1C1) produced a maximum output of 125 μW.

When the number of anodes remained single, and the number of cathodes was increased to two (A1C2), the maximum output changed only 4.8% compared to A1C1 (A1C2: 131 μW). This could be attributed to the fact that when the number of anodes was not increased, the number of electrons received from the microbes did not increase either, so the output did not increase proportionally.

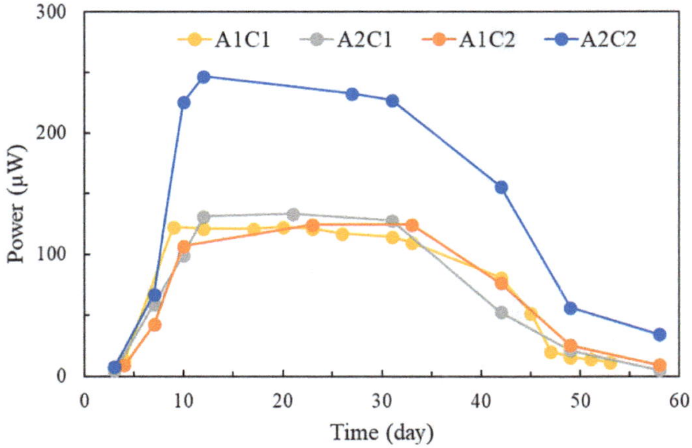

Fig. 3 Four types of FMFC outputs: (**a**) one anode and one cathode (A1C1), (**b**) two anodes and one cathode (A2C1), (**c**) one anode and two cathodes (A1C2), and (**d**) two anodes and two cathodes (A2C2)

On the other hand, when the number of anodes was increased to two (A2C1) while the number of cathodes remained one, the maximum power changed by only 0.8% compared to A1C1 (A2C1: 124 μW). This is thought to be because even though the number of anodes increased and the number of electrons supplied by the microbes increased, there is a limit to the number of reduction reactions that can occur on one cathode.

These results indicate that if the number of electrodes is increased to improve the output of the FMFC, both the anodes and the cathodes must be increased by the same number.

Increasing both the anode and cathode to two (A2C2) resulted in a maximum power of 1.96 times that of A1C1 (A2C2: 246 μW). This result suggests that when the number of anodes and cathodes are the same, the number of electrodes and the output power obtained are almost proportional to each other.

3.2 Scaled-Up FMFC Output Evaluation

Figure 4 shows the power of the FMFC with the number of electrodes scaled up to four. The output power of the FMFC with the anodes set horizontally was 144 μW, 164 μW, and 467 μW on the 4th, 8th, and 14th days of operation, respectively. This was 4.9, 4.9, and 4.2 times the output of the single-electrode FMFC, respectively.

Power generation continued until the amount of organic matter in the solution in which the FMFC floated decreased enough to affect microbial metabolism (52 days).

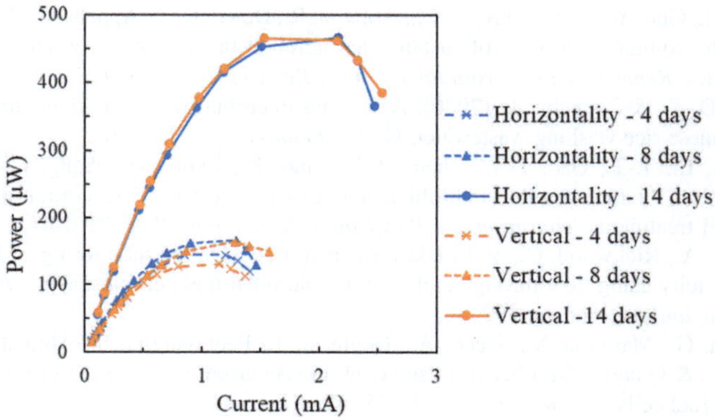

Fig. 4 Power curves of a four-electrode FMFC with horizontal or vertical anodes on the 4th, 8th, and 14th days of operation

Even when the anodes were set up in the vertical direction, no change in output was observed compared to the horizontal case. Thus, the vertical direction is considered to be more suitable for practical use, as multiple anodes can be compactly accommodated.

4 Conclusion

In order to facilitate the practical application of FMFC, this study focused on the number of block-shaped electrodes to be scaled up. Increasing either the anodes or the cathodes did not improve the output of the FMFC. In other words, scale-up requires increasing both anodes and cathodes.

The FMFC, which was scaled up with both anodes and cathodes to four, produced a maximum output of 467 μW. This is 4.2 times the output of the single-electrode FMFC, so even with four electrodes, there was a proportional relationship between the number of electrodes and output power.

It would be worthwhile to increase the number of electrodes further and proceed with scale-up.

References

1. Ma, J., Zhang, J., Zhang, Y., Guo, Q., Hu, T., Xiao, H., Lu, W., & Jia, J. (2023). Progress on anodic modification materials and future development directions in microbial fuel cells. *Journal of Power Sources, 556*, 232486.
2. Yaqoob, A. A., Al-Zaqri, N., Alamzeb, M., Hussain, F., Oh, S.-E., & Umar, K. (2023). Bioenergy generation and phenol degradation through microbial fuel cells energized by domestic organic waste. *Molecules, 28*(11), 4349.

3. Kamali, M., Guo, Y., Aminabhavi, T. M., Abbassi, R., Dewil, R., & Appels, L. (2023). Pathway towards the commercialization of sustainable microbial fuel cell-based wastewater treatment technologies. *Renewable and Sustainable Energy Reviews, 173*, 113095.
4. Nguyen, D.-T., & Taguchi, K. (2020). A floating microbial fuel cell: Generating electricity from Japanese rice washing wastewater. *Energy Reports, 6*(9), 758–762.
5. Pu, K.-B., Li, T.-T., Gao, J.-Y., Chen, Q.-Y., Guo, K., Zhou, M., Wang, C.-T., & Wang, Y.-H. (2022). Floating flexible microbial fuel cells for electricity generation and municipal wastewater treatment. *Separation and Purification Technology, 300*, 121915.
6. Adekunle, A., Rickwood, C., & Tartakovsky, B. (2020). Online monitoring of heavy metal–related toxicity using flow-through and floating microbial fuel cell biosensors. *Environmental Monitoring and Assessment, 192*, 52.
7. Massaglia, G., Margaria, V., Sacco, A., Tommasi, T., Pentassuglia, S., Ahmed, D., Mo, R., Pirri, C. F., & Quaglio, M. (2018). In situ continuous current production from marine floating microbial fuel cells. *Applied Energy, 230*, 78–85.
8. Hirose, S., Nguyen, D. T., & Taguchi, K. (2022). Proposal of a block-shaped rice husk–bokuju cathode with added copper in a floating microbial fuel cell. *Energy Reports, 8*(15), 841–847.
9. Martinez, S. M., & Lorenzo, M. D. (2019). Electricity generation from untreated fresh digestate with a cost-effective array of floating microbial fuel cells. *Chemical Engineering Science, 198*, 108–116.
10. Tatinclaux, M., Gregoire, K., Leininger, A., Biffinger, J. C., Tender, L., Ramirez, M., Torrents, A., & Kjellerup, B. V. (2018). Electricity generation from wastewater using a floating air cathode microbial fuel cell. *Water-Energy Nexus, 1*(2), 97–103.

Exploring Liquefied Dimethyl Ether for Lipid Extraction from Fat Balls in Wastewater Pumping Stations

Febrian Rizkianto, Kazuyuki Oshita, Ryosuke Homma, and Masaki Takaoka

1 Introduction

Fat, oil, and grease (FOG) that are discharged into the wastewater can accumulate in the sewer system [1]. They can form solid deposits through the saponification process between calcium, free fatty acids, and other debris in the wastewater [2]. Pumping stations are susceptible to FOG accumulation, often resulting in blockages of the wastewater flow. These FOG deposits, known as fat balls, tend to accumulate on the water surface of the pumping station [1]. Without proper management of FOG disposal, the expected increase in FOG deposits in the near future is a cause for concern. Therefore, it is crucial to identify a feasible approach for the utilization of this waste. Despite its negative effect, previous studies have demonstrated the potential of fat balls as an alternative lipid source for biodiesel, with a lipid content of approximately 93% on a dry basis [3]. This could help eliminate disposal costs and promote increased renewable energy production.

Typically, wastewater lipids are extracted using an organic solvent [4]. However, this process is energy-intensive, and the management of solvent residue poses challenges. As an alternative, dimethyl ether (DME) has been promoted to extract neutral and complex lipids [5, 6]. DME exhibits partial miscibility with water (7–8 wt.%) and high affinity toward organic compounds, enabling the processing of wet biomass [7]. Additionally, it can be liquefied (at 0.51–0.59 MPa) and degassed by adjusting the pressure [8], thereby offering a cost-effective and energy-efficient solution [9].

The present work aims to investigate the recovery of lipids from fat balls using dimethyl ether extraction. The objective of this study is to investigate the recovery

F. Rizkianto · K. Oshita (✉) · R. Homma · M. Takaoka
Department of Environmental Engineering, Graduate School of Engineering, Kyoto University, Kyoto, Japan
e-mail: oshita.kazuyuki.6e@kyoto-u.ac.jp

P. Pong (ed.), *Renewable Energy Resources and Conservation*, Green Energy and Technology, https://doi.org/10.1007/978-3-031-59005-4_19

of lipids from fat balls using liquefied DME (L-DME) extraction. The efficiency of oil recovery using L-DME extraction is compared to that of mechanical shaking extraction with hexane. The properties of recovered lipids were compared, including the elemental composition, functional groups, and fatty acids characterization.

2 Materials and Methods

2.1 Materials Collection

In this experiment, fat balls were collected from the water surface of a pumping station in Kobe City (Japan), which receives 37,000 m^3 of wastewater per day. The collected samples were stored at 4 °C before use and homogenized before experiments. Fat balls collected from the pumping station had a solid texture and were found attached to light-floating substances. Table 1 presents the characteristics of fat balls.

2.2 Analytical Methods

The total solid and volatile solids were determined using the standard method 2540 G [10]. The carbon (C), hydrogen (H), and nitrogen (N) contents were measured using a C/H/N analyzer (JM10; J-Science Lab Co., Ltd., Kyoto, Japan). The calorific value was determined using a bomb calorimeter (CA-4J; Shimadzu Co. Ltd., Kyoto, Japan). The lower heating value (LHV) was calculated from the measured higher heating values (HHV) [11]. The functional group composition was measured by Fourier transform infrared (FTIR) spectroscopy in ATR mode (IRSpirit-T; Shimadzu Co. Ltd., Kyoto, Japan).

The extracted lipids were converted into fatty acid methyl esters using a mixture of sulfuric acid and methanol (1% v/v) (Guaranteed Reagent, Wako Co., Ltd., Japan) and heated in a constant temperature bath (EYELA Co., Ltd., Tokyo, Japan) at

Table 1 Characteristics of fat balls

Parameter	Fat balls	Unit
Total solids (TS)	55.1	%
Volatile matter (VS)	92.4	%TS
C	68.4	%
H	11.4	%
N	0.2	%
O	12.5	%
HHV	39.16	MJ/kg
LHVw	19.02	MJ/kg

50 °C overnight. The profiles of methylated fatty acids were analyzed using Gas Chromatography-Mass Spectrometry (GC-MS) (Shimadzu Co. Ltd., Kyoto, Japan) with a capillary column [100 m × 0.20 μm × 0.25 μm].

2.3 *Lipid Extraction by Mechanical Shaking with Hexane and Liquefied DME*

The lipid extraction experiments using hexane were conducted in a mechanical shaker with a ratio of 2:1 (hexane/fat balls) for 60 min at ambient temperature (150 rpm). After shaking, the tube was centrifuged at 3000 rpm for 10 min. The supernatant phase was collected and dried. The remaining solid content represents the oil content. For the lipid extraction with liquefied dimethyl ether (L-DME), the apparatus was arranged in series, as depicted in Fig. 1. L-DME was generated by cooling pure gaseous DME (Tamiya Co., Ltd, Japan) to –12 °C with ethanol and ice, and then stored in vessel 1 (100 cm^3). The fat balls (1 g) and glass beads were loaded in vessel 2 (10 cm^3). The DME/sample ratio is 100 mL/g, extraction pressure was set at 0.7 MPa, and the flow rate was maintained at 10 cm^3/min. Vessel 1 was immersed in a water bath, and the temperature was maintained at 37 °C. Subsequently, L-DME was transferred from vessel 1 to vessel 2 under pressurization at 0.7 MPa. After extraction at room temperature, the lipids in vessel 3 (100 cm^3) were recovered by depressurizing the L-DME to 0.1 MPa, enabling complete evaporation of the DME. Subsequently, the liquid mixture was filtered and collected for analysis.

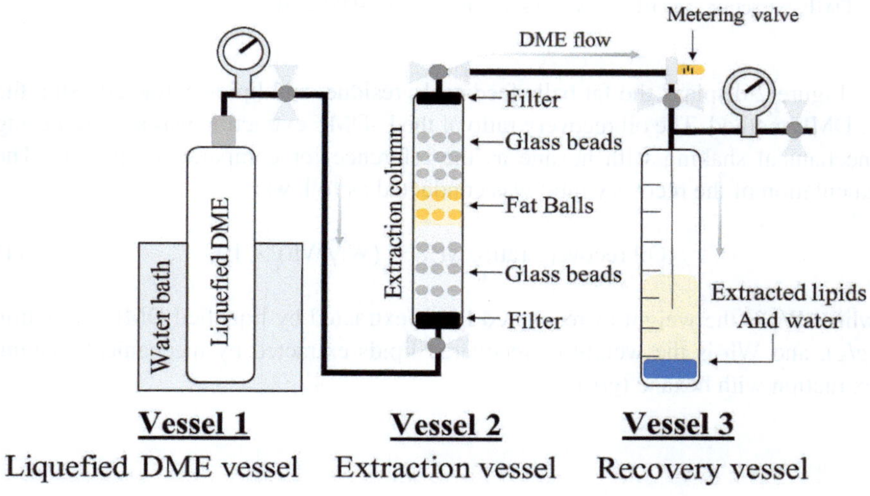

Fig. 1 DME equipment apparatus

Fig. 2 Photo images of (**a**) original fat balls, (**b**) homogenized fat balls, (**c**) fat balls residue after L- DME extraction, and (**d**) recovered substances after L-DME extraction

Figure 2 displays the fat balls feedstock, residue, and lipids extracted using the L-DME method. The oil recovery ratio of the L-DME extraction was assessed using mechanical shaking with hexane as the reference for comparative analysis. The calculation of the recovery ratio was conducted as follows:

$$\text{Oil recovery ratio, wt\%} = (\text{Wr}/\text{Wh}) \times 100 \quad (1)$$

where Wr is the weight of recovered lipids extracted by liquefied DME extraction (g/g), and Wh is the weight of recovered lipids extracted by mechanical shaking extraction with hexane (g/g).

3 Result and Discussions

3.1 Lipid Extraction Yield, Elemental Analysis, and Mass Balance

The oil recovery ratio of L-DME extraction, compared to hexane extraction, was 89% (Table 2). This indicates that lipid extraction using L-DME achieved a slightly different performance compared to the solvent extraction method. The C/H/N composition of the recovered oil was also identical. The elemental composition of lipids was 71.2–72.3% carbon, 12.6–12.7% hydrogen, and 0.13–0.18% nitrogen. Figure 3 illustrates the mass balances before and after the DME treatment. The mass of solids and water content were significantly reduced to approximately 0.3 g and 0.02 g, respectively.

Table 2 Extraction yield, recovery ratio, and characteristics of lipids

Extraction methods	Lipid, Yield, %	Oil recovery ratio[a], %		C, %	H, %	N, %
L-DME extraction	46.0	89.0	Extracted liquid	71.2	12.7	0.18
			Solid residue	67.0	11.6	0.58
Hexane extraction	52.0		Extracted liquid	72.3	12.6	0.13
			Solid residue	68.1	11.4	0.52

[a]Based on the yield of lipid by hexane extraction method

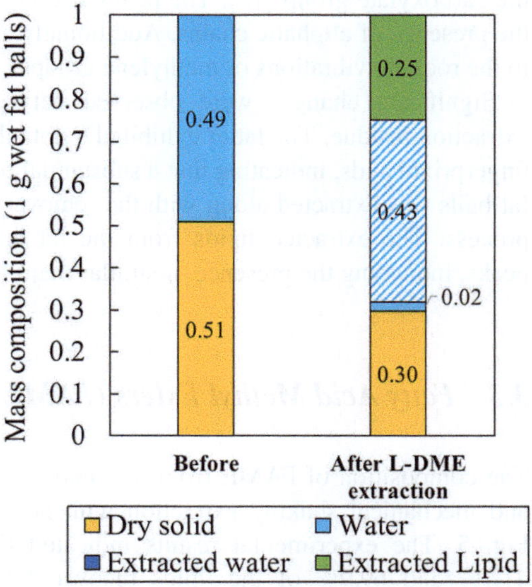

Fig. 3 Mass balance before and after extraction of lipids from fat balls

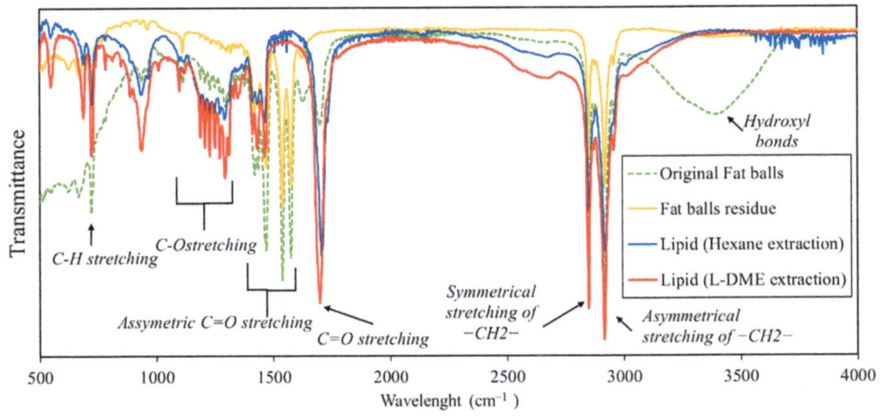

Fig. 4 FTIR spectra of original fat balls, residue, and extracted lipids

3.2 FTIR Analysis

Figure 4 illustrates the FTIR spectra of the original fat balls, the fat balls residue after extraction, and the recovered lipids. The bands observed at 3400 cm^{-1} correspond to hydroxyl bonds [12]. The bands detected at 2850–2955 cm^{-1} indicated the presence of abundant saturated hydrocarbons [9]. The peaks around 1740 cm^{-1} are attributed to the C=O stretching vibration of esters of lipids. The bands at 1530–1610 cm^{-1} were assigned to the asymmetric stretching vibration of the carboxylate group [13]. The peaks ranging from 1350 to 1180 cm^{-1} represent the presence of aliphatic chains. Additionally, the band at 720 cm^{-1} was attributed to the rocking vibrations of methylene groups.

Significant changes were observed between the original fat balls and the extraction residue. The latter exhibited a notable decrease in functional groups and fingerprint bands, indicating that a substantial portion of the lipids from the original fat balls was extracted along with the removal of water during the lipid extraction process. The extracted lipids from the fat balls displayed similar transmittance peaks, indicating the presence of similar major components.

3.3 Fatty Acid Methyl Esters (FAME) Analysis of Lipids

The composition of FAME from the lipids extracted from fat balls using L-DME and mechanical shaking extraction with hexane was compared, as presented in Fig. 5. The experimental results indicated that C18 fatty acids accounted for 41.7% and 69.9% of the lipids recovered by L-DME and hexane extraction, respectively. Furthermore, it was observed that the recovered oil obtained through

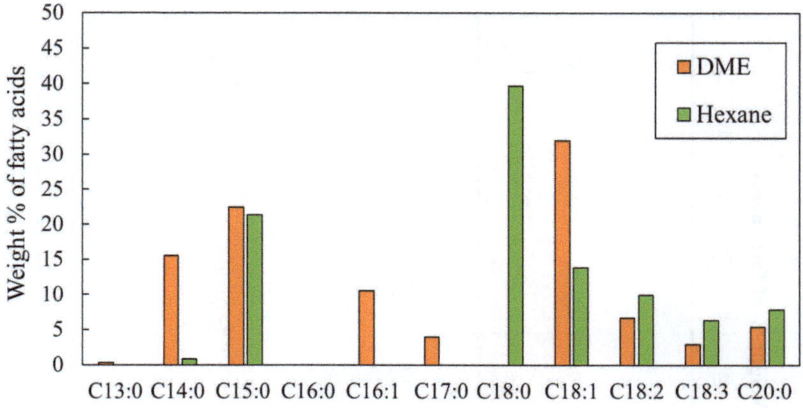

Fig. 5 Fatty acids composition of methyl esters derived from fat balls

DME extraction contained higher proportions of C13–14 (38.3%) and C16 (10.6%) fractions.

The proportion of unsaturated fatty acids (UFA) in the recovered oil by DME extraction was higher (52%) compared to the hexane extraction method (30.2%) (Fig. 6). High content of unsaturated fatty acids (UFA) in wastewater lipids has been reported by previous studies [14]. The predominant presence of UFAs is highly advantageous in terms of biodiesel's oxidative stability, particularly in colder regions [10]. Overall, the results demonstrate that the lipids recovered from fat balls can be extracted using both DME extraction and hexane extraction, resulting in differences in the composition of fatty acids.

4 Conclusion

This study investigated the recovery of lipids from fat balls using the liquefied DME extraction method, and the results were compared to the mechanical shaking extraction method using hexane. The lipid recovery using DME extraction (46%) was slightly lower compared to the mechanical shaking extraction (52%). However, it is noteworthy that the lipids recovered through DME extraction exhibited a higher content of unsaturated fatty acids, which are known to enhance the quality of produced biodiesel. Further research is required to explore potential modifications of the extraction parameters in the DME extraction, as such modifications have the potential to improve the quality of both lipids and methyl esters.

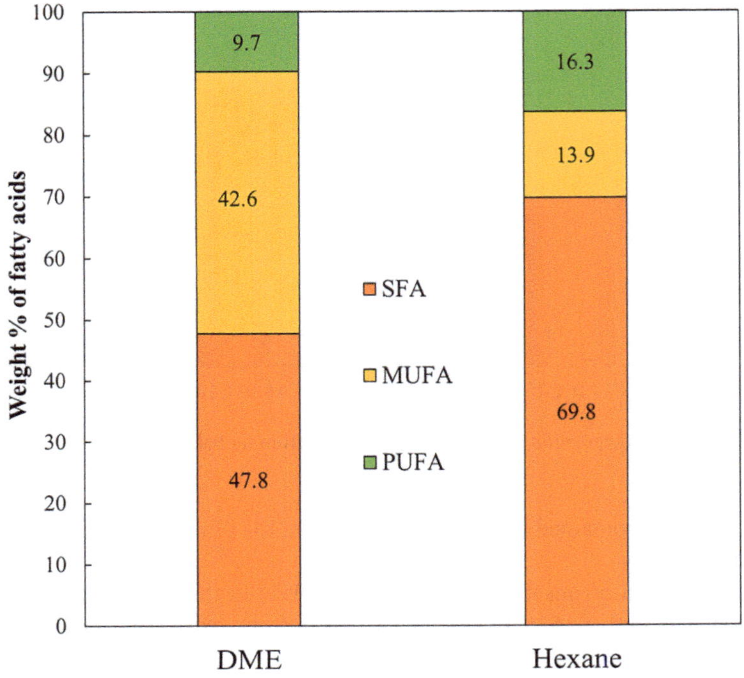

Fig. 6 Fatty acid profiles of methyl esters according to saturation level

References

1. Williams, J. B., Clarkson, C., Mant, C., Drinkwater, A., & May, E. (2012). Fat, oil and grease deposits in sewers: Characterisation of deposits and formation mechanisms. *Water Research*, 46, 6319.
2. He, X., de los Reyes, F. L., & Ducoste, J. J. (2012). A critical review of fat, oil, and grease (FOG) in sewer collection systems: Challenges and control. *Critical Reviews in Environmental Science and Technology, 47*, 1191.
3. Collin, T., Cunningham, R., Jefferson, B., & Villa, R. (2020). Characterisation and energy assessment of fats, oils and greases (FOG) waste at catchment level. *Waste Management, 103*, 399.
4. Frkova, Z., Venditti, S., Herr, P., & Hansen, J. (2020). Assessment of the production of biodiesel from urban wastewater-derived lipids. *Resources, Conservation and Recycling, 162*, 105044.
5. Bauer, M. C., & Kruse, A. (2019). The use of dimethyl ether as an organic extraction solvent for biomass applications in future biorefineries: A user-oriented review. *Fuel, 254*, 115703.
6. Kanda, H., Li, P., Ikehara, T., & Yasumoto-Hirose, M. (2012). Lipids extracted from several species of natural blue-green microalgae by dimethyl ether: Extraction yield and properties. *Fuel, 95*, 88.
7. Kanda, H., Fukuta, Y., Wahyudiono, & Goto, M. (2021). Enhancement of lipid extraction from soya bean by addition of dimethyl ether as entrainer into supercritical carbon dioxide. *Foods, 10*, 1223.
8. Oshita, K., Toda, S., Takaoka, M., Kanda, H., Fujimori, T., Matsukawa, K., & Fujiwara, T. (2015). Solid fuel production from cattle manure by dewatering using liquefied dimethyl ether. *Fuel, 159*, 7.

9. Zhang, D., Huang, Y., Oshita, K., Takaoka, M., Ying, M., Sun, Z., & Sheng, C. (2021). Crude oil recovery from oily sludge using liquefied dimethyl ether extraction: A comparison with conventional extraction methods. *Energy and Fuels, 35*, 17810.
10. Olkiewicz, M., Fortuny, A., Stüber, F., Fabregat, A., Font, J., & Bengoa, C. (2015). Effects of pre-treatments on the lipid extraction and biodiesel production from municipal WWTP sludge. *Fuel, 141*, 250.
11. Chen, M., Oshita, K., Mahzoun, Y., Takaoka, M., Fukutani, S., & Shiota, K. (2021). Survey of elemental composition in dewatered sludge in Japan. *Science of the Total Environment, 752*, 141857.
12. Iasmin, M., Dean, L. O., Lappi, S. E., & Ducoste, J. J. (2014). Factors that influence properties of FOG deposits and their formation in sewer collection systems. *Water Research, 49*, 92.
13. He, X., de los Reyes, F. L., Leming, M. L., Dean, L. O., Lappi, S. E., & Ducoste, J. J. (2013). Mechanisms of Fat, Oil and Grease (FOG) deposit formation in sewer lines. *Water Research, 47*, 4451.
14. Vieira Magalhães-Ghiotto, G. A., Sílvio, M. P. M., Trevisan, E., & Arroyo, P. A. (2022). Extraction and characterization of the lipids from domestic sewage sludge and in situ synthesis of methyl esters. *Environmental Progress & Sustainable Energy*, e14027.

Generation of Electrical Energy Through Microbial Fuel Cells Using Beet Waste As Fuel

Rojas-Flores Segundo, Santiago M. Benites, De La Cruz-Noriega Magaly, Nazario-Naveda Renny, Nélida Milly Otiniano, and Daniel Delfín-Narciso

1 Introduction

The dependence on fossil sources for the generation of electrical energy in recent times has increased intensely due to the expansion of electrification to rural areas, but it is still insufficient due to the high costs that this means [1, 2]. There are reports that in 1978 energy consumption was 270.5 EJ and in 2018 it was 580 EJ worldwide, according to BP Statistical Review of Eorl Energy (June 2019) [3], increasing CO2 levels and harming the environment [4]. Reason why the use of renewable energies opens as a sustainable alternative for these towns or camps far from the big cities. It is estimated that the production of electricity of this type will increase by 350% by the year 2030 [5, 6]. Within these energies is the energy from biomass, where microbial fuel cells are a promising technology because they use different types of waste to generate electricity [6]. There are several types of designs in microbial fuel cells (MFCs), being single-chamber microbial fuel cells (MFCs-SC) the most promising due to their low manufacturing cost as they do not use external aeration for their operation [7, 8]. All MFCs are made up of two chambers (cathodic and anodic) where these chambers are almost always joined by a proton

R.-F. Segundo (✉) · S. M. Benites · D. L. C.-N. Magaly · N.-N. Renny
Vicerrectorado de Investigación, Universidad Autónoma del Perú, Lima, Peru
e-mail: santiago.benites@autonoma.pe; mdelacruzn@autonoma.edu.pe; renny.nazario@autonoma.pe

N. M. Otiniano
Instituto de Investigación en Ciencias y Tecnología de la Universidad Cesar Vallejo, Trujillo, Peru
e-mail: notiniano@ucv.edu.pe

D. Delfín-Narciso
Grupo de Investigación en Ciencias Aplicadas y Nuevas Tecnologías, Universidad Privada del Norte, Trujillo, Peru
e-mail: daniel.delfin@upn.edu.pe

P. Pong (ed.), *Renewable Energy Resources and Conservation*, Green Energy and Technology, https://doi.org/10.1007/978-3-031-59005-4_20

exchange membrane on the inside and an external circuit on the outside. One of the great advantages of using this technology is that any type of fuel (substrate) can be used to generate electricity, due to the oxidation-reduction process that converts chemical energy into electricity [9–11].

On the other hand, organic waste has also become a big problem, because governments, mainly in developing countries, do not have an adequate collection center for this type of garbage [12]; reason why the merchants of the supply centers throw their waste in the streets surrounding them, thus generating contamination for the neighbors [13]. It has been reported that in 2018 organic waste was 1.3 billion tons and that by 2025 it will be 2.2 billion, one of the most thrown wastes is vegetables [14]. In this sense, beetroot is a vegetable widely used in people's daily diets, due to its components such as pectin, lignin, hemicelluloses, glucose, cellulose, mannose, and arabinose [15]; which bring great health benefits; In addition, the Food Agriculture Organization (FAO) (2017) mentioned that 86% of this type of waste is reused in cellulose-based microfibers, which produce stable liquids for use in detergents or paints; in arabinose, which has demonstrable health benefits, it is also used as galacturonic acid which is a main constituent of pectin, used for polymers [16, 17]. But so far, its use as fuel for electric power generation has not been reported, other types of vegetables or fruits have been found in the literature as fuel in MFCs, for example, Javed et al. (2021) used vegetable waste adjusted to a pH of 4.5 as fuel in their MFCs, managing to generate voltage peaks and power density of 550 m v and 81.12 W/m^2 using graphite rods as electrodes [18]. Likewise, Kalagbor et al. (2023) used fruit waste as fuel in a single-bed MFCs on a medium scale, managing to generate voltage peaks and electric current of 4.2 V and 3.7 mA using graphite electrodes [19]. While Mulyono et al. (2022) used mustard and spinach waste as a substrate, managing to generate voltage peaks of 0.804 V and an electric current of 2.37 mA [20].

The main objective of this research work is to observe the potential of beet waste for the generation of electrical energy using single-chamber microbial fuel cells as an electronic device. For this, the values of voltage, pH, electrical conductivity, and electrical current will be obtained, which will be monitored for 30 days. Also, the values of electrical resistance, power density, and electrical current density will be calculated. This research will give a second promising use to this type of waste to use it as fuel to improve people's quality of life through the supply of electrical energy in an environmentally sustainable manner.

2 Materials and Methods

(a) Single chamber fuel cell design: The design used was based on the foundation of single-chamber fuel cells (MFCs-SC), which were manufactured in triplicate. The MFCs-SC from the company SAIDKOCC (Fujian, China) with a volume of 100 mL were used, while those used were copper (Cu, area = 40 cm^2) and zinc (Zn, area = 62.5 cm^2) as anode and cathode, respectively. The electrodes were connected on the outside by means of a circuit with an external resistance of 100 Ω and on the inside a proton exchange membrane (Nafion 117, Merck) was used.

(b) Selection and preparation of beet waste: Beet waste was used as a substrate, which was collected from the La Hermelinda market, Trujillo, Peru; 1 kg was collected which was taken to the laboratory to be washed to eliminate any type of impurity acquired from the environment. The waste was crushed in an extractor (Labtron, LDO-B10-USA) obtaining approximately 900 mL.

(c) Instruments used for the characterization of MFCs-SC: To measure electrical parameters such as voltage and electrical current, a multimeter (Prasek Premium PR-85, USA) with an external resistance of 100 Ω was used, while for the measurement of the ph and electrical conductivity a pH-meter (110 Series Oakton, USA) and a conductivity meter (CD-4301, USA) were used. The values of the internal resistance were gauged using a Vernier energy sensor (Vernier-± 30 V & ± 1000 mA, USA), while for the values of power density and current density, the method and resistances were used in the same way as that used by Rojas-Flores et al. (2022) [21].

3 Results and Analysis

The MFCs-SC showed a peak voltage value of 1.03 ± 0.25 V on the eleventh day and then decreased until the last day (0.29 ± 0.21 V), as can be seen in Fig. 1a. Successive increase of the voltage values is due to the initial oxidation-reduction reactions that occurred in the anodic chamber, which create a potential differential between both electrodes, whose potential differential decays in the last one to the corrosion that was observed in the cathode electrode [22, 23]. While the electric current values showed their peak value on the eleventh day (5.64 ± 0.75 mA) they increased by 355.32% with respect to their initial value (0.63 ± 0.04 mA) on the first day. The initial increase in these electric current values is due to good biofilm formation on the anode electrode, and to the ideal conditions for microorganisms to proliferate, while these values decreased due to fermentation and sedimentation observed in the last days of monitoring [23–25]. In Fig. 1c the variation of the pH values shown during the monitoring is observed, observing a pH of 4.86 ± 0.34 as the optimum on day 11; these pH values varied over time, due to the fact that they were not used a stabilizer and biological and chemical processes caused this value to change [26, 27]. The values of the electrical conductivity of the beet waste varied from day 1, showing a value of 175.71 ± 6.87 mS/cm on day 11, and then slowly decayed, as observed in Fig. 1d; the increase in this value is due to the release of electrons by the organic matter present in the substrate during the process of generating electrical energy [28].

Figure 2a shows the average of the internal resistance of the MFCs-SC, which was calculated by Ohm's Law, where the voltage data was placed on the "*Y*" axis and the electric current on the "*X*" axis, whose slope of the linear adjustment

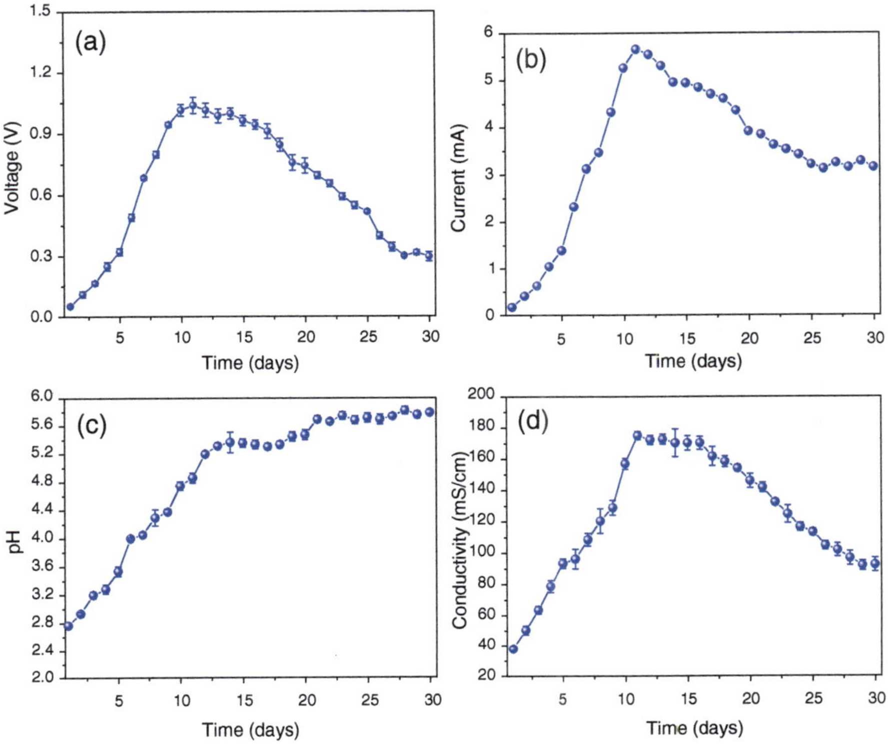

Fig. 1 Values obtained from (**a**) voltage, (**b**) electrical current, (**c**) pH, and (**d**) electrical conductivity obtained from monitoring microbial fuel cells

represents the internal resistance, yielding a value of 48.253 ± 4.749 Ω. According to the literature, a good adhesion of the biofilm with the electrode and a high electrical conductivity would be the main reasons for such a low internal resistance found, as well as the use of metallic electrodes opposing a low resistance to the passage of electrons [29, 30]. The power density values as a function of the current density are shown in Fig. 2b, where the maximum power density was calculated at 96.741 ± 4.874 mW/m^2 at a current density of 5.614 mA/m, with a peak voltage of 925.64 ± 8.64 V. Although the power density values are not higher than those of other investigations, this investigation shows high values of voltage and electric current, this may be due to the fact that the corrosion shown in the final days affected the efficiency of the electronic device, for which reason it is recommended to cover the electrodes with some type of substance compatible with microorganisms. In Fig. 3, the generation of electrical energy is observed connecting the three MFCs-SC connected in series, managing to turn on an LED light (green).

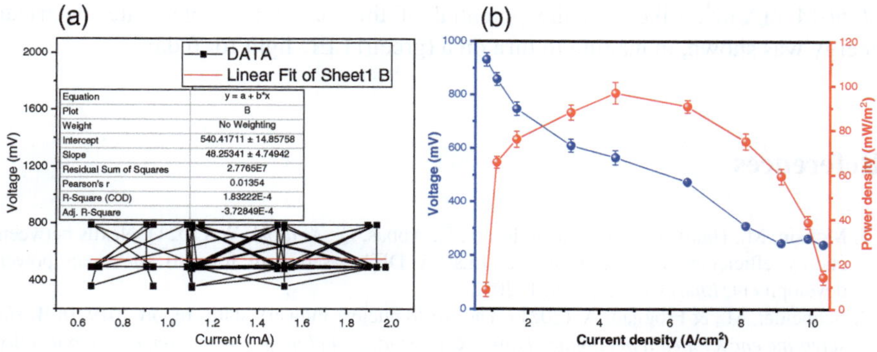

Fig. 2 (**a**) Internal resistance and (**b**) power density as a function of the current density of the MFCs-SC

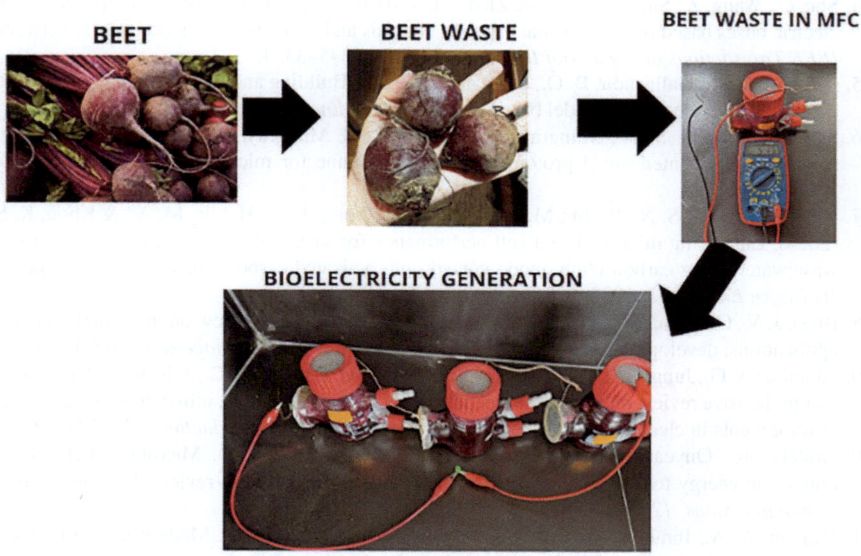

Fig. 3 Bioelectricity generation process through MFCs-SC connected in series

4 Conclusions

It was possible to successfully show the potential of single-chamber microbial fuel cells for the generation of electrical energy using beet waste as fuel. The monitoring carried out for 30 days showed maximum values of 1.03 ± 0.25 V and 5.64 ± 0.75 mA of voltage and electric current, which operated on the eleventh day at an optimum pH of 4.86 ± 0.34 with 175.71 ± 6.87 mS/cm of electric conductivity. Likewise, a low internal resistance of 48,253 ± 4749 Ω was observed, whose maximum power density was 96,741 ± 4874 mW/m^2 with a current density of 5614 mA/m^2. Likewise, the potential of this substrate to generate electrical energy was shown, managing to turn on a (green) LED light for 6 days.

References

1. Mohsin, M., Hanif, I., Taghizadeh-Hesary, F., Abbas, Q., & Iqbal, W. (2021). Nexus between energy efficiency and electricity reforms: A DEA-based way forward for clean power development. *Energy Policy, 149*, 112052.
2. Schneider, M., & Froggatt, A. (2021). The world nuclear industry status report 2019. In *World scientific encyclopedia of climate change: Case studies of climate risk, action, and opportunity* (Vol. 2, pp. 203–209). World Scientific.
3. Rahman, M. M. (2020). Environmental degradation: The role of electricity consumption, economic growth and globalisation. *Journal of Environmental Management, 253*, 109742.
4. She, C., Wang, Z., Sun, F., Liu, P., & Zhang, L. (2019). Battery aging assessment for real-world electric buses based on incremental capacity analysis and radial basis function neural network. *IEEE Transactions on Industrial Informatics, 16*(5), 3345–3354.
5. Brinkerink, M., Gallachóir, B. Ó., & Deane, P. (2021). Building and calibrating a country-level detailed global electricity model based on public data. *Energy Strategy Reviews, 33*, 100592.
6. Surti, P., Kailasa, S. K., Mungray, A., Park, T. J., & Mungray, A. K. (2024). Vermiculite nanosheet augmented novel proton exchange membrane for microbial fuel cell. *Fuel, 357*, 130046.
7. Ng, C. A., Chew, S. N., Bashir, M. J., Abunada, Z., Wong, J. W., Habila, M. A., & Khoo, K. S. (2024). Enhancing microbial fuel cell performance for sustainable treatment of palm oil mill wastewater using carbon cloth anode coated with activated carbon. *International Journal of Hydrogen Energy, 52*, 1092–1104.
8. Boas, J. V., Oliveira, V. B., Simões, M., & Pinto, A. M. (2022). Review on microbial fuel cells applications, developments and costs. *Journal of Environmental Management, 307*, 114525.
9. Palanisamy, G., Jung, H. Y., Sadhasivam, T., Kurkuri, M. D., Kim, S. C., & Roh, S. H. (2019). A comprehensive review on microbial fuel cell technologies: Processes, utilization, and advanced developments in electrodes and membranes. *Journal of Cleaner Production, 221*, 598–621.
10. Obileke, K., Onyeaka, H., Meyer, E. L., & Nwokolo, N. (2021). Microbial fuel cells, a renewable energy technology for bio-electricity generation: A mini-review. *Electrochemistry Communications, 125*, 107003.
11. Yaqoob, A. A., Ibrahim, M. N. M., & Guerrero-Barajas, C. (2021). Modern trend of anodes in microbial fuel cells (MFCs): An overview. *Environmental Technology & Innovation, 23*, 101579.
12. Abad, V., Avila, R., Vicent, T., & Font, X. (2019). Promoting circular economy in the surroundings of an organic fraction of municipal solid waste anaerobic digestion treatment plant: Biogas production impact and economic factors. *Bioresource Technology, 283*, 10–17.
13. Ashokkumar, V., Flora, G., Venkatkarthick, R., SenthilKannan, K., Kuppam, C., Stephy, G. M., et al. (2022). Advanced technologies on the sustainable approaches for conversion of organic waste to valuable bioproducts: Emerging circular bioeconomy perspective. *Fuel, 324*, 124313.
14. Paes, L. A. B., Bezerra, B. S., Deus, R. M., Jugend, D., & Battistelle, R. A. G. (2019). Organic solid waste management in a circular economy perspective–A systematic review and SWOT analysis. *Journal of Cleaner Production, 239*, 118086.
15. Stevanato, P., Chiodi, C., Broccanello, C., Concheri, G., Biancardi, E., Pavli, O., & Skaracis, G. (2019). Sustainability of the sugar beet crop. *Sugar Tech, 21*, 703–716.
16. Usmani, Z., Sharma, M., Diwan, D., Tripathi, M., Whale, E., Jayakody, L. N., et al. (2022). Valorization of sugar beet pulp to value-added products: A review. *Bioresource Technology, 346*, 126580.

17. Rana, A. K., Gupta, V. K., Newbold, J., Roberts, D., Rees, R. M., Krishnamurthy, S., & Thakur, V. K. (2022). Sugar beet pulp: Resurgence and trailblazing journey towards a circular bioeconomy. *Fuel, 312*, 122953.
18. Javed, M. M., Nisar, M. A., & Ahmad, M. U. (2021). Effect of NaCl and pH on bioelectricity production from vegetable waste extract supplemented with cane molasses in dual chamber microbial fuel cell. *Pakistan Journal of Zoology, 54*(1), 247–254.
19. Kalagbor, I. A., Azunda, B. I., Igwe, B. C., & Akpan, B. J. (2020). Electricity generation from waste tomatoes, banana, pineapple fruits and peels using single chamber microbial fuel cells (SMFC). *Journal of Waste Management & Xenobiotics, 3*, 000142.
20. Mulyono, T., Misto, M., Cahyono, B. E., & Fahmidia, N. H. (2022, September). The impact of adding vegetable waste on the functioning of microbial fuel cell. In *AIP conference proceedings* (Vol. 2663, No. 1, p. 020008). AIP Publishing LLC.
21. Rojas-Flores, S., Nazario-Naveda, R., Benites, S. M., Gallozzo-Cardenas, M., Delfín-Narciso, D., & Díaz, F. (2022). Use of pineapple waste as fuel in microbial fuel cell for the generation of bioelectricity. *Molecules, 27*(21), 7389.
22. Borja-Maldonado, F., & Zavala, M. Á. L. (2022). Contribution of configurations, electrode and membrane materials, electron transfer mechanisms, and cost of components on the current and future development of microbial fuel cells. *Heliyon, 8*, e09849.
23. Itoshiro, R., Yoshida, N., Yagi, T., Kakihana, Y., & Higa, M. (2022). Effect of ion selectivity on current production in sewage microbial fuel cell separators. *Membranes, 12*(2), 183.
24. Rojas-Flores, S., De La Cruz-Noriega, M., Nazario-Naveda, R., Benites, S. M., Delfín-Narciso, D., Rojas-Villacorta, W., & Romero, C. V. (2022). Bioelectricity through microbial fuel cells using avocado waste. *Energy Reports, 8*, 376–382.
25. Yaqoob, A. A., Al-Zaqri, N., Yaakop, A. S., & Umar, K. (2022). Potato waste as an effective source of electron generation and bioremediation of pollutant through benthic microbial fuel cell. *Sustainable Energy Technologies and Assessments, 53*, 102560.
26. Wang, Y., Zhang, X., & Lin, H. (2024). Effects of pH on simultaneous Cr (VI) and p-chlorophenol removal and electrochemical performance in Leersia hexandra constructed wetland-microbial fuel cell. *Environmental Technology, 45*, 483–494.
27. Littfinski, T., Beckmann, J., Gehring, T., Stricker, M., Nettmann, E., Krimmler, S., et al. (2022). Model-based identification of biological and pH gradient driven removal pathways of total ammonia nitrogen in single-chamber microbial fuel cells. *Chemical Engineering Journal, 431*, 133987.
28. Wilberforce, T., Abdelkareem, M. A., Elsaid, K., Olabi, A. G., & Sayed, E. T. (2022). Role of carbon-based nanomaterials in improving the performance of microbial fuel cells. *Energy, 240*, 122478.
29. ElMekawy, A., Hegab, H. M., Dominguez-Benetton, X., & Pant, D. (2013). Internal resistance of microfluidic microbial fuel cell: Challenges and potential opportunities. *Bioresource Technology, 142*, 672–682.
30. Liang, P., Huang, X., Fan, M. Z., Cao, X. X., & Wang, C. (2007). Composition and distribution of internal resistance in three types of microbial fuel cells. *Applied Microbiology and Biotechnology, 77*(3), 551–558.

Part IV
Waste Heat Utilization and Energy Conservation

Effect of a Movable Phase Change Materials (PCMs) Layer on Lowering Energy Usage in Desert Structures

Maamar Hamdani, Ayoub Aggoune, Yacine Marif, Sidi Mohammed El Amine Bekkouche, Saleh Al-Saadi, Mohamed Kamal Cherier, and Rachid Djeffal

1 Introduction

During the past decade, the building sector in Algeria has emerged as one of the major consumers of energy, accounting for approximately 40% of total fossil energy consumption [1, 2]. According to the annual statistics on energy consumption in residential buildings provided by the National Energy Balance Sheet of the Ministry of Energy, the residential sector in Algeria has witnessed a steady increase, with a share of 40% in 2010, 43% in 2013, and 46.7% in 2020 [2]. Additionally, Algeria is particularly vulnerable to extreme heat, with the Algerian desert experiencing record-breaking temperatures, especially in southern regions such as Illizi, Ouargla, Adrar, and Ghardaia, where temperatures reached 51 degrees Celsius in July 2021 [1]. This exceptional situation has resulted in a substantial increase in energy consumption, imposing a heavy burden on the economy. Moreover, there have been challenges in implementing regulations and decisions related to energy conservation [2]. As a result, current policies are focused on improving building energy performance and promoting the use of renewable energy sources. An effective strategy to enhance energy efficiency in the construction sector is to prioritise the building envelope, which plays a crucial role in energy loss or gain depending on its structure and orientation [3, 4]. It has become necessary

M. Hamdani (✉) · S. M. E. A. Bekkouche · M. K. Cherier · R. Djeffal
Unité de Recherche Appliquée en Energies Renouvelables, URAER, Centre de Développement des Energies Renouvelables, CDER, Ghardaia, Algeria

A. Aggoune · Y. Marif
Department of Physics, Faculty of Mathematics & Matter Sciences, University of Ouargla, LENREZA Laboratory, Ouargla, Algeria

S. Al-Saadi
Department of Civil and Architectural Engineering, Sultan Qaboos University, Muscat, Oman

P. Pong (ed.), *Renewable Energy Resources and Conservation*, Green Energy and Technology, https://doi.org/10.1007/978-3-031-59005-4_21

to ensure that the building envelope meets the minimum requirements outlined in the National Building Energy Code, as this serves as an indicator for achieving occupant comfort and reducing the energy consumption of heating and cooling systems [1]. One interesting approach to reducing energy demands is to focus on the design of earth-sheltered traditional buildings. These buildings have the potential to significantly reduce energy consumption and outperform modern exposed buildings that utilise new materials without proper consideration of their suitability for hot climate regions. The lack of proper consideration has resulted in a heavy reliance on conventional air conditioning systems to ensure thermal comfort. To enhance energy efficiency in the construction sector, prioritising the building envelope is an effective strategy. The envelope's structure and orientation play a crucial role in energy loss or gain. Meeting the minimum requirements outlined in the National Building Energy Code is essential to occupant comfort and reducing energy consumption in heating and cooling systems [1]. Scholars have extensively researched both active and passive energy-saving strategies to reduce building energy consumption while maintaining indoor thermal comfort [5–7]. Active technologies often face challenges related to low energy efficiency [8], leading to increased interest in passive energy-saving technologies [9]. These passive strategies encompass various techniques such as integrating Trombe walls [10, 11], lightweight concrete walls [12], insulation materials [13], retro-reflective materials [14], green roofs [15], and phase-change materials (PCMs) [16, 17]. Incorporating PCM into building envelopes is to be encouraged because of its thermal capacity and excellent latent heat. The use of PCM for air conditioning and heating buildings has given rise to much research in recent decades. A large number of contributions have been published on the application of these materials as passive systems. However, the choice of the melting temperature in accordance with different climatic conditions plays a key role in improving the energy performance of buildings [2]. Servando et al. [4] has shown that the use of PCMs coupled with night ventilation is a very powerful strategy to reduce the cooling demand of buildings. Osterman et al. [5] proposed work on the use of filled composite wall MCPs, and the results showed that integrating MCPs into walls can reduce daily energy consumption. Zhou et al. [6] had used MCP sheets as the interior cladding of the walls and ceilings in a south-facing room of an office building in Beijing. Qingang Xiong et al. [7] used the technique of trombe walls with phase change materials (PCM) to tackle the problem of energy consumption to heat and cool buildings. Various designs of trombe walls and different types of phase change materials are presented. Numerical modelling has been widely employed to assess the performance of phase-change materials (PCMs) in building applications [11, 14, 15]. In a study conducted in a hot climate region of South Algeria, the energy performance of building envelopes integrated with PCMs for cooling load reduction was evaluated using Trnsys 18. Parametric studies were conducted to investigate factors such as phase change temperature (ranging from 22 to 32 °C), PCM location (exterior versus interior), and PCM thickness (ranging from 3 to 20 mm). The findings suggested that PCMs applied to the exterior surfaces of walls exhibited better performance, while thinner PCM layers demonstrated higher efficiency and cost benefits. Another study by Soares et al. [16] utilised a combination of EnergyPlus and GenOpt tools to perform a multidimensional optimisation study on PCM-integrated drywall in a lightweight steel-framed residential building located in different climates. The researchers concluded that for each climate, an optimal solution incorporating PCM-drywall can be achieved, leading to significant annual energy savings in heating and cooling. PCM-drywalls were particularly suitable for warmer climates. Additionally, Chan [17] evaluated the energy performance of a typical residential flat integrated with PCM-integrated walls in Hong Kong using EnergyPlus. I also described in my own post the prudent method of incorporating phase change materials into the building envelope in a desert climate, when considering the integration of conventional PCMs into the building envelope that resulted in a 36.4% decrease in annual energy consumption [2]. and when integrated PCMs meticulously, a 50% cost reduction is realised.

2 Methodology

2.1 Description of the Simulated Building

The model for our study is an apartment with four facades. The construction is located on an area of 82.8 m^2 according to Fig. 1. The proposed construction is composed of walls, wooden doors, and glass windows, with a reinforced concrete floor and slab. The exterior walls consist of several layers of building materials: hollow brick (20 cm), cement mortar, and plaster. Conversely, the interior walls (partitions) are 15 cm thick in brick plus a layer of plaster coating. The building has a main door on the north side and has large windows of 1.44 m^2 each on the east and west sides, respectively, making it exposed to sunlight and natural ventilation from all sides [2]. The facades of this apartment are subdivided into two types: one is main and includes some openings, and the other is secondary and does not carry any openings. The height of the walls is 2.8 m. This model consists of two bedrooms with an area of 10.8 m^2 for the first and 12 m^2 for the second, a living room of 10.8 m^2, a kitchen, and a bathroom [2]. In our study, we chose the ceiling of the living room as a witness to the integration of the PCM to weaken or even eliminate the effects of the thermal load of the envelope. PCMs can also be passively incorporated into ceilings on a permanent basis. In this case, we put the PCMs in the inner layer in the form of lightweight, portable sheets that can be removed and replaced whenever needed, according to the study [2].

Our idea was to see the effect of PCMs on the evolution of the internal temperatures of the different zones of the habitat under the climatic demands of an arid climate, and this throughout the year, with priority given to days when the outside temperature converges with the melting point of the materials with phase change. Taking advantage of the PCM's ability to store and release heat was our goal in this work. The simulations are based on the building envelope reinforcement by using a layer of removable phase change materials, using PCM panels on the internal

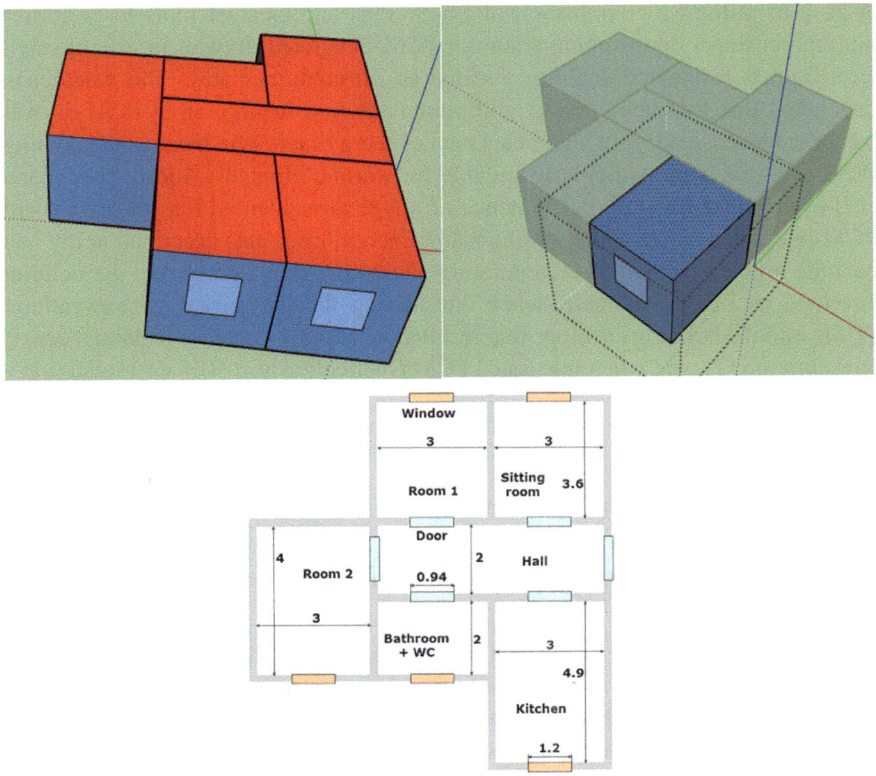

Fig. 1 Floor plan, 3D sketch

ceiling of the sitting room, and comparing it with the same room without PCM. We then analyse the monthly energy savings and the average indoor air temperatures of the room by modelling the thermal and energy performance of a building by TRNSYS-18 and type (285) [2], and through the results obtained, we can make special configurations.

2.2 *Weather and Climate Data for Ghardaia City*

Algeria is the largest country in Africa and the tenth largest in the globe (2,381,741 square kilometres). The Sahara encompasses roughly 90% of Algeria's landmass. Algeria has numerous climate zones. To accomplish this, we utilise the Copen Geiger world map [15, 18], which is a reference and is constantly updated, as it provides a classification of climates based on rainfall and temperature and indicates the existence of seven climatic zones distributed from relatively wet to very dry

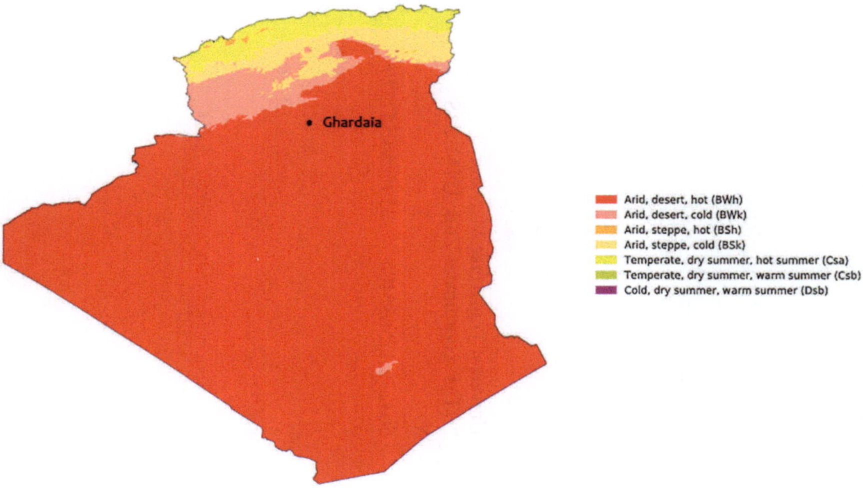

Fig. 2 Ghardaia in its geographic location and the Koppen-Geiger climate classification map for Algeria

(desert-like), with the desert climate comprising the majority. Using the Koppen-Geiger classification, we constructed a map of productive climate zones. Thus, the barren climate type (B) covers approximately 95% of the country's total land area, while the mild temperate climate type (C) covers only about 5%. The most prevalent climate zone for the arid (B) climate type is the hot desert type (BWh), which encompasses more than 85% of the country, followed by the frigid steppe climate zone (BSk), which covers about 5%. Almost exclusively, Type C climate zones are characterised by the entirely humid mild temperate climate with hot summer (Cfa) climate zone. Notable is the fact that, with the exception of the BWh zone, all other climate zones are exclusive to northern Algeria. On the Algerian map, Fig. 2 depicts the location of Ghardaia, which serves as the entrance to the Algerian Sahara [18, 19].

Ghardaia is located in the northern-central Saharan region, about 600 km to the south of the Algerian capital (Algiers). The detailed coordinates are summarised in Table 1.

The region has a desert climate, characterised by lengthy, scorching summers and mild winters with brief, heated days and chilly nights. This climate continues to be dominated by heat, drought, and large diurnal and annual variations in temperature. The sunshine duration determined by Yaiche et al. [19, 20], for the period between 1992 and 2002, revealed that the annual average daily sunshine duration for the Ghardaia region ranges from 9 to 9 h and 30 min. As shown in Fig. 3, these established distributions are firmly linked to the average cloudiness (cloud cover) of the sites.

Table 1 Ghardaïa coordinates

Geographical coordinates			Coordinates in degrees and decimal minutes		Universal Transverse Mercator "UTM" coordinates Zone UTM: 31S	
Latitude	Longitude	Altitude from sea level	Latitude	Longitude	*X*	*Y*
32.4909400°	3.6734700°	503 m	32°29.4564′ Nord	3°40.4082′ Est	563272.17563548	3595054.5575299

Ghardaïa, Algérie: faits et chiffres, 2021

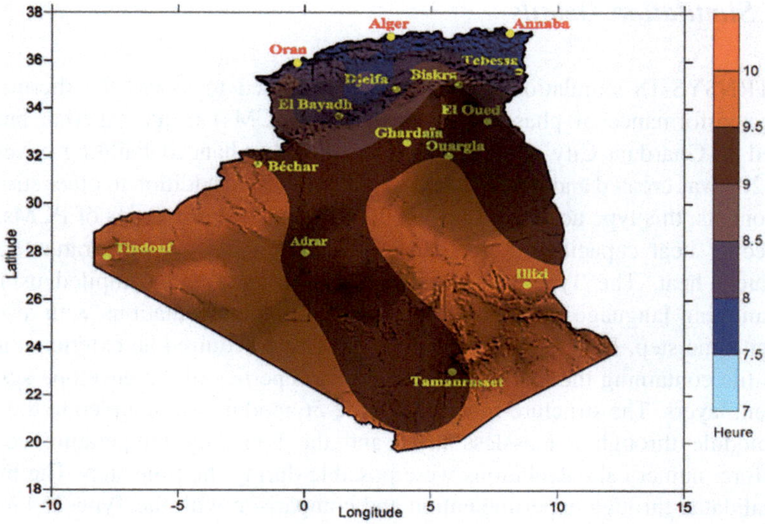

Fig. 3 Mapping of annual mean values of measured sunshine duration, from 1992 to 2002 [19]

Table 2 Climate weather data for Ghardaïa [2]

	Mean daily max. temp. [C]	Mean daily min. temp. [C]	Mean daily temp. [C]	Relative humidity [%]	Global solar radiation [MJ/m^2]	Mean wind speed [m/s]
January	17.24	3.0	11.83	57.06	3341.02	3.09
February	17.76	2.7	12.30	43.33	4383.26	3.94
March	23.30	3.2	17.66	32.91	5886.47	4.18
April	28.11	7.0	21.93	28.81	6834.47	4.21
May	32.27	14.0	26.42	27.19	7611.29	3.82
June	38.04	22.0	31.64	24.23	7878.23	4.09
July	41.29	24.0	35.18	24.23	7876.09	2.17
August	40.28	21.0	33.96	27.71	7192.66	3.19
September	34.04	15.5	28.48	41.27	6055.46	3.16
October	28.84	13.8	23.49	40.26	4774.48	3.53
November	21.27	6.3	16.28	51.13	3683.42	3.71
December	17.27	3.6	11.99	59.40	3025.16	3.13

Source of climate datasets

Solar irradiation received by a flat surface is estimated to range between 2.35 and 6.86 kWh/m^2 per day, with an annual mean of 4.98 kWh/m^2 per day. This potential is thus quite substantial and can be exploited practically throughout the year. According to Table 1, additional bibliographic references (sources of climate datasets) have reported significant climatic conditions for Ghardaia (Table 2).

2.3 Simulation Details

The TRNSYS-18 simulation> environment was used to assess the thermal and energy performance of phase-change materials (PCMs) in an Algerian building located in Ghardaia City. In order to model PCM-enhanced building envelopes, Type-285 was created and incorporated into TRNSYS. In addition to other structural components, this type accounted for the thermo-physical properties of PCMs, such as specific heat capacities, layer densities, melting points, temperature ranges, and latent heat. The Type-285 module was developed and compiled using the programming language FORTRAN in order to enable calculations with a wholly implicit time step. Integration of the new PCM type required an external configuration file containing the thermal and physical properties of the envelope system's different layers. The structure-modelling Type-56 module was coupled to the Type-285 module through a massless layer and the boundary temperature concept. Therefore, numerical calculations were possible during the time step. The module was validated through experimentation and comparison with the Type-260 module developed by Kuznik et al. (2010). In the simulation, the efficacy of PCM ceiling panels was evaluated by comparing the indoor thermal environment and energy consumption of PCMs with varying melting temperatures to those of a room without PCM under the same boundary conditions [20]. A parametric study was conducted, varying only the PCM's melting temperature while considering environmental conditions and investigating various natural ventilation scenarios. Several factors, including temperature requirements and thermal, physical, chemical, and economic properties that contribute to energy efficiency, influence the selection of a phase-change material for building applications [21–23]. The PCM panels were installed within the interior of the roofing layer, as this was deemed the optimal location for replacement ease and thermal comfort in the living area. The panels had ventilation holes on multiple surfaces to effectively regulate and stabilise the ambient temperature. Ceilings with detachable PCM panels help maintain a specific temperature by absorbing and storing heat during warmer periods and recycling it later. This leads to a more consistent interior temperature and lower energy costs [21, 22]. When the temperature during the summer exceeds a certain threshold, the air conditioning is activated to provide comfort. However, if PCM materials are incorporated into the ceiling, air conditioning can be deferred at night, thereby reducing energy consumption. Using PCMs, the energy required to cool the home is stored and not lost, thereby maintaining a predetermined temperature throughout the day and night. This improves energy efficiency and thermal comfort. The objective of this study was to compare the thermal comfort differences between two different commercial paraffin PCMs, RT21 and RT26, placed within removable ceiling panels and exposed to hot, mild, and frigid conditions. The results demonstrated that installing PCMs on the interior side of the ceiling roof partition effectively reduced internal temperature variations during mild-to-cold and temperate weather. During mildly heated conditions, the PCM within the partition of the PCM ceiling panels minimised surface temperature fluctuations [2].

3 Results and Discussion

The results demonstrated that the installation of PCM in the buildings results in a decrease in peak cooling and heating demand. The energy demand was reduced by as much as 10.15% when using PCM RT21, and when using PCM RT26, we get a reduction of 17.03%, but with the PCM removable panels, we have 21.31%. These ratios are very effective, especially since we used them only on the ceiling of the room, as demonstrated in Fig. 4.

The results indicate a decrease in the percentage of energy consumption in March, April, and May, where the lowest numbers were recorded at 25.7 kW, 27.8 kW, and 114 kW, respectively, without placing phase-changing materials inside the building. The month of June also witnessed the beginning of a rise in consumption. It reaches its peak consumption in July with an amount of 480.3 kW, and it continues to rise in the month of August at 312.2 kW. In this case, the most suitable solution is to put on PCM Rt21 in both June and July, but in the month of August, we have replaced the PCM with another type, PCM Rt26, because it is the most suitable. The results showed a remarkable decrease in temperature with and without PCMs. This is what makes us think before using it throughout the year, and in some months of the year it must be removed because it does an undesirable reverse action, which increases the demand for energy. Then we put on PCM Rt21 in September, October, November, December, and January because it gives good results compared to PCM Rt26. In February, the PCM Rt26 gives satisfactory results.

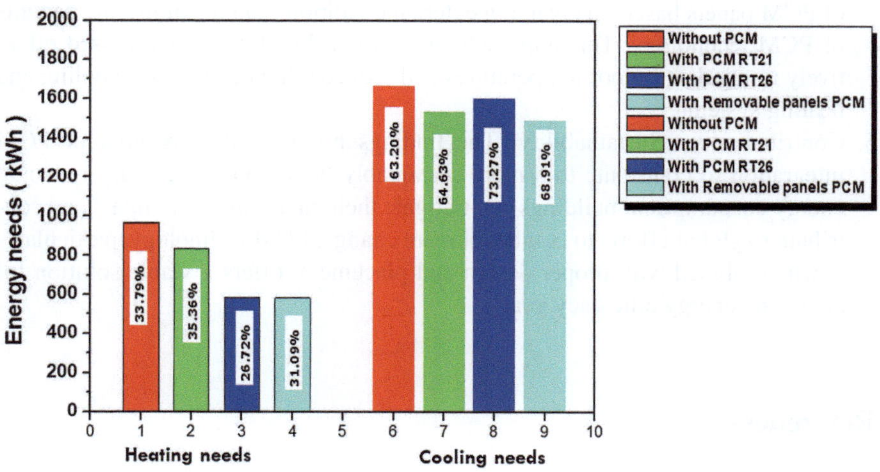

Fig. 4 Energy consumption in the heating and cooling period due to the conventional and removable panel integration of PCM

4 Conclusions

Based on the research conducted on the application of phase change materials (PCMs) in the building envelope in Ghardaia city, South Algeria, the following conclusions and recommendations can be drawn:

1. Conventional PCM Integration: Integrating fixed PCM RT21 in the ceiling of the building envelope resulted in a 10.15% reduction in annual energy consumption. Replacing it with a fixed PCM RT21 throughout the year further increased the energy savings to 17.03%. This highlights the effectiveness of PCM integration in reducing energy consumption.
2. Removable PCM Panels: The study revealed the potential for even greater energy savings by using removable PCM panels strategically positioned within the ceiling, considering seasonal variations. This approach resulted in an impressive reduction of 28.68% in annual energy consumption. It demonstrates the importance of adjusting PCM placement based on changing seasons to optimise energy efficiency.
3. Significance of Proper PCM Integration: The research emphasises the significance of proper PCM integration in the building envelope. Selecting the appropriate PCM type and considering its placement can significantly contribute to energy savings and improved building performance. It is crucial to carefully analyse the specific requirements and conditions of the building to determine the most effective PCM integration strategy.
4. Consideration of Seasonal Variations: Taking seasonal variations into account is essential for optimising energy efficiency. Adapting the placement and utilisation of PCM panels based on changing external conditions can maximise the benefits of PCM technology. This approach ensures that PCM systems are used effectively to regulate indoor temperatures and reduce reliance on active cooling and heating systems.
5. Contribution to Sustainability: The findings highlight the potential of PCM integration to contribute to a more sustainably built environment. By reducing energy consumption, buildings can mitigate their environmental impact and contribute to global efforts to combat climate change. PCM technology, particularly when combined with proper design and placement, offers a viable solution for achieving energy efficiency goals.

References

1. Algérie Presse Service. (2019). *Le secteur du bâtiment, premier consommateur d'énergie en Algérie*. http://www.aps.dz/economie/85470-le-secteur-du-batiment-premier-consommateurd-energie-en-algerie. Accessed January, 2021.
2. Maamar, H., Bekkouche, S. M. E. A., Al-Saadi, S., KamalCherier, M., Djeffal, R., & Zaiani, M. (2021). Judicious method of integrating phase change materials into a building envelope under Saharan climate. *International Journal of Energy Research, 45*, 18048.

3. Hamdani, M., Bekkouche, S. M. A., Benouaz, T., & Cherier, M. K. (2015). A new modelling approach of a multizone building to assess the influence of building orientation in Saharan climate. *Thermal Science, 19*(Suppl. 2), S591–S601.
4. Álvareza, S., Cabezab, L. F., Ruiz-Pardoa, A., Castell, A., & Tenorioc, J. A. (2013). Building integration of PCM for natural cooling of buildings. *Applied Energy, 109*, 514–522.
5. Osterman, E., Butala, V., & Stritih, U. (2015). PCM thermal storage system for 'free' heating and cooling of buildings. *Energy and Buildings, 106*, 125–133.
6. Zhou, G., Yang, Y., Wang, X., & Zhou, S. (2009). Numerical analysis of effect of shape-stabilized phase change material plates in a building combined with night ventilation. *Applied Energy, 86*(1), 52–59.
7. Xiong, Q., Alshehri, H. M., Monfaredi, R., Tayebi, T., Majdoub, F., Hajjar, A., Delpisheh, M., & Izadi, M. (2022). Application of phase change material in improving trombe wall efficiency: An up-to-date and comprehensive overview. *Energy and Buildings, 258*, 111824.
8. Mathis, D. (2019). *Développement de nouveaux matériaux de haute inertie thermique à base de bois et matériaux à changement de phase biosourcés* (PhD Thesis). Laval University, Canada.
9. Delcroix, B. (2015). *Modeling of thermal mass energy storage in buildings with Phase Change Materials* (PhD Thesis). Mechanical Engineering Department Polytechnic School of Montréal, Canada.
10. Aketouane, Z., Malha, M., Bruneau, D., et al. (2018). Energy savings potential by integrating phase change material into hollow bricks: The case of Moroccan buildings. *Building Simulation, 11*, 1109–1122.
11. Jin, X., Medina, M. A., & Zhang, X. (2016). Numerical analysis for the optimal location of a thin PCM layer in frame walls. *Applied Thermal Engineering, 103*, 1057–1063.
12. Wang, J., Long, E., Qin, W., & Xu, L. (2013). Ultrathin envelope thermal performance improvement of prefab house by integrating with phase change material. *Energy and Buildings, 67*(210), 216.
13. Mi, X., Liu, R., Cui, H., Memon, S. A., Xing, F., & Lo, Y. (2016). Energy and economic analysis of building integrated with PCM in different cities of China. *Applied Energy, 175*(1), 324–336.
14. Louanate, A., El Otmani, R., Kandoussi, K., & Boutaous, M.'. H. (2020). Dynamic modeling and performance assessment of single and double phase change material layer–integrated buildings in Mediterranean climate zone. *Journal of Building Physics*. https://doi.org/10.1177/1744259120945
15. Rubel, F., & Kottek, M. (2010). Observed and projected climate shifts 1901–2100 depicted by world maps of the Köppen-Geiger climate classification. *Meteorologische Zeitschrift, 19*(2), 135–141. https://doi.org/10.1127/0941-2948/2010/0430
16. Xiong, Q., Alshehri, H. M., Rezvan, M., Tayebi, T., Majdoub, F., Hajjar, A., Delpisheh, M., & Zadi, M. (2022). Application of phase change material in improving trombe wall efficiency: An up-to-date and comprehensive overview. *Energy and Buildings, 258*(6), 111824. https://doi.org/10.1016/j.enbuild.2021.111824
17. Djeffal, R., Mohamed, K. C., Bekkouche, S. M. E. A., Younsi, Z., Hamdani, M., & Al-Saadi, S. (2022). Concept development and experimentation of a Phase Change Material (PCM) enhanced domestic hot water. *Journal of Energy Storage, 51*, 104400. https://doi.org/10.1016/j.est.2022.104400
18. Zeroual, A., Assani, A. A., Mohamed, M., & Alkama, R. (2018). Assessment of climate change in Algeria from 1951 to 2098 using the Köppen–Geiger climate classification scheme. *Climate Dynamics, 52*, 227.
19. Yaiche, M. R., Bouhanik, A., Bekkouche, S. M. A., Malek, A., & Benouaz, T. (2014). Revised solar maps of Algeria based on sunshine duration. *Energy Conversion and Management, 82*, 114–123.
20. Soares, N., Gaspar, A. R., Santos, P., & Costa, J. J. (2014). Multi-dimensional optimization of the incorporation of PCM-drywalls in lightweight steel-framed residential buildings in different climates. *Energy and Buildings, 70*, 411.

21. Yang, R., Li, D., Arıcı, M., Salazar, S. L., Zhang, C., & Fu, Q. (2023). Thermal performance of an innovative double-skin ventilated façade with multistep-encapsulated PCM integration. *Journal of Energy Storage, 73*, 109121.
22. Allam, B., Nehari, T., & Benlekkam, M. L. (2023). Building brick wall thermal management optimization and temperature control based on phase change materials integration. Case study of the city of Bechar, Algeria. *Journal of Energy Storage, 73*, 109043.
23. Al-Yasiri, Q., & Szabó, M. (2023). Numerical analysis of thin building envelope-integrated phase change material towards energy-efficient buildings in severe hot location. *Sustainable Cities and Society, 89*, 104365.

Energy Optimization Analysis and Case Study of Commercial Buildings Using EnergyPlus

Qitong Huang

1 Introduction

Energy is a vital resource for human civilization, but over-reliance on fossil fuels has caused environmental issues and increased demand for renewable energy. Cities consume a large portion of global energy and emit greenhouse gases. Therefore, reducing energy waste, especially in commercial buildings, is crucial.

Commercial building energy consumption varies by type, with healthcare buildings accounting for a significant portion due to their large floorspace and growing energy demand. Hospitals have an average of 264,800 sf per building compared to outpatient health care buildings of 13,600 sf [1]. Healthcare buildings account for the largest energy consumption among all commercial buildings, with Shanghai hospitals consuming twice that of public buildings [2].

Architects need to predict energy consumption in the early design stage. It is because optimizing building design from an energy sustainability perspective is essential. Heating, cooling, lighting, and ventilation are dominant energy uses in buildings, and selecting suitable shapes and orientations can reduce energy consumption by 30–40% [3]. Healthcare buildings, which consume more energy than other commercial buildings, can benefit from optimal building design at the conceptual design stage, which will significantly reduce total life cycle costs.

To evaluate healthcare buildings' energy performance, EnergyPlus was used as a simulation tool. It includes integrated simulation, multi-zone air flow, HVAC loops, and algorithms from the new ASHRAE loads toolkit. Hourly energy simulations were performed to test the energy models' efficiency and performance for the aim of reducing the life cycle cost of buildings.

Q. Huang (✉)
Southwest Jiaotong University-Leeds Joint School, Chengdu, Sichuan, China

P. Pong (ed.), *Renewable Energy Resources and Conservation*, Green Energy and Technology, https://doi.org/10.1007/978-3-031-59005-4_22

2 Literature Review

This section gives a comprehensive literature review about building energy consumption, current energy consumption status of healthcare buildings, the comparison of EnergyPlus, and other energy simulation tools. The reason why EnergyPlus has been chosen in this chapter and the importance optimizing building design at design stage was stated as well.

2.1 Building Energy Consumption

The energy crisis has been exacerbated by the rapid growth of the global population and urbanization, leading to increased energy consumption. Buildings, in particular, have played a significant role in this trend. The size of buildings increased faster than before with the former rising by 56% from 3.8 million to 5.9 million while the latter grew from 51 billion to 97 billion square feet [1]. Energy is essential for economic development, but it is also a finite resource that is becoming increasingly scarce. Commercial buildings account for over 30% of the total energy consumption of civil buildings and have a large service sector that consumes several energy services [2]. Unfortunately, due to greenhouse gas emissions, the formation of urban heat island (UHI) may have a significant negative impact on building energy consumption by increasing space cooling demand and reducing space heating demand [4]. Reducing energy consumption in commercial buildings is necessary to address the energy crisis. This can be achieved through initial design optimization of building systems [3]. Additionally, increasing the proportion of sustainable energy sources is also crucial in ensuring sufficient energy for the future, the advancements in net-sero energy buildings (NZEBs) method considering energy infrastructure connections, renewable energy sources, and energy-efficiency measures also reduce energy consumption and greenhouse gas [5].

2.2 Energy Consumption of Healthcare Building

Healthcare buildings are big energy consumers. The average floor area of it exceeds that of commercial buildings. Consuming a lot of energy is due to special requirements such as strict air quality and lighting system requirements in buildings, electricity demand for special medical equipment, 24-h and 365-day healthcare activities, and energy consumption of HVAC systems. Compared to other ordinary commercial buildings, healthcare buildings, especially those with the largest floor area, have become one of the most energy-efficient buildings. The unit energy consumption of healthcare buildings is about twice that of ordinary commercial buildings and 10–20 times that of residential buildings. The main departments of a hospital require different equipment or systems to achieve their functions, including air conditioning and lighting systems, medical equipment, heating systems, and elevators. This makes the energy use of hospital buildings very complex, with numerous types of energy, a wide range of fields involved, and different energy usage times. Studying and mastering the spatiotemporal distribution of hospital energy consumption is key to hospital energy conservation. HVAC accounts for a large proportion of energy consumption in hospital buildings, making electricity and fuel consumption seasonal. An energy consumption audit of a large hospital building in Shanghai found that annual electricity consumption follows a parabolic distribution, with the highest electricity consumption in August and the largest proportion of air conditioning energy consumption in building electricity consumption, with the highest in August. Hospital HVAC system comprehensive service system, lighting power consumption, various medical equipment, electric heaters, elevator power consumption; Heating, domestic hot water, processing, and cooking consume natural gas [2].

2.3 Development of Energy Simulation Tools

Various building energy modeling tools have been developed, including EnergyPlus, DOE, TRANSYS, and ESP-r. EnergyPlus is the most popular due to its cost-effectiveness, rapid release, and ability to be easily modified and extended using Fortran90. The program uses integrated simulation to avoid inaccurate space temperature prediction. It includes three basic programs: simulation manager, heat and mass balance simulation module, and building systems simulation module. The simulation manager controls all simulation loops and is designed to address issues of spaghetti code and lack of structure in previous programs. The heat and mass balance manager contacts the building systems simulation manager, which controls the simulation of HVAC and electrical systems, equipment, and components. EnergyPlus also includes a complex CFD simulation for fluid movement, which meets the fundamental assumption of the heat balance model. The building systems simulation manager is designed to meet the requirement of realistically limited capacity and tight connection of air and water side of the system and plant. EnergyPlus is a current option for simulating energy consumption in buildings [6].

2.4 Design Stage of Buildings

According to Keoleian and Menerey, 90% of the building life cycle cost is determined during the design phase. Initial design optimization can reduce life cycle costs [7]. Nearly 53.2% of building energy consumption is used for space heating. Altan Dombayci proposed a new insulation method that reduces energy consumption by 46.6% and reduces environmental impact compared to traditional methods [8]. Dad-Khoung Bui et al. adopted a computational optimization method based on the adaptive facade system design to adjust its heat and visible perspective rate according to dynamically changing climatic conditions, reducing the energy consumption of the two case studies by at least 14%, respectively [9]. Improving the control of HVAC systems is the key to reducing building energy consumption [3]. Early prediction of energy consumption is very important for energy conservation and emission reduction work. Intelligent building control can reduce energy consumption by 50% −80% −60% in space heating, cooling, lighting, and ventilation areas [8]. Architects need to focus more on equipment life cycle operations and long-term returns rather than initial investments [10].

3 Case Studies of Building Energy Simulation Using EnergyPlus

The US Department of Energy (DOE) collaborated with three of its national laboratories to develop commercial reference buildings, which were formerly known as commercial building benchmark models. These reference buildings play a crucial role in the program's energy modeling software research by providing comprehensive descriptions for whole building energy analysis using EnergyPlus simulation software. There are 16 building types that represent approximately 70% of the commercial buildings in the USA. DOE developed 16 reference building types that represent most commercial buildings across 16 locations, which encompass all US climate zones [1]. In this chapter, hospital and outpatient building types are chosen for case studies. For both hospital and outpatient building types, three different design parameters are examined: energy consumption changes in buildings, standing direction, temperature setpoint, and construction material. These three variables are chosen because they can be artificially controlled and have a significant impact on the energy performance of the building.

3.1 Hospital Building Type

A five-story hospital in Miami, Florida has been chosen to examine the effects of varying design specifications on building energy consumption. The hospital's exterior is shown in Fig. 1. Three design parameters will be varied: standing direction, temperature setpoint, and construction material. These variations will be studied to observe monthly and annual variations in energy consumption and the proportion of each energy source consumed, with the goal of identifying the most effective strategies for reducing energy consumption. Standing direction refers to the direction in which the building is facing, which is northward in this case. Temperature setpoint pertains to the desired cooling temperature in patient rooms.

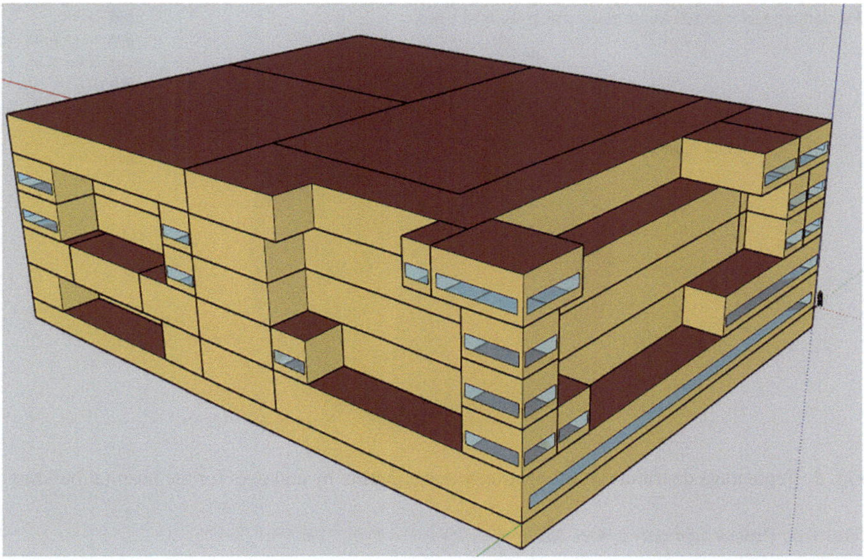

Fig. 1 The hospital building viewed in Google Sketchup

Below is the summary of the features of this hospital:

- Latitude and longitude are 25.78 and −85.27, respectively, and the building is located in a time zone of −5 h with an elevation of 2 m
- It has 55 zones, including patient rooms, faculties rooms, and so on
- Standing direction is 0° to the north
- For patients' rooms, the default heating temperature set-point is 21.1 °C and cooling set-point is 22.2 °C
- Conductivity of the 8 inch concrete wall is 1.311 W/m.K.

To analyze the impact of design parameters on building energy consumption, it is important to first determine the percentage of each item's consumption as a proportion of total electricity consumption. In this case, there are a total of eight categories that have been summarized:

1. Cooling
2. Interior Equipment
3. Interior Lights
4. Fans
5. Exterior Equipment
6. Pumps
7. Heat Rejection: Electricity
8. Others

As shown in Fig. 2, the percentage distribution of electricity consumption by end uses for the hospital building is displayed. It can be observed that except for the

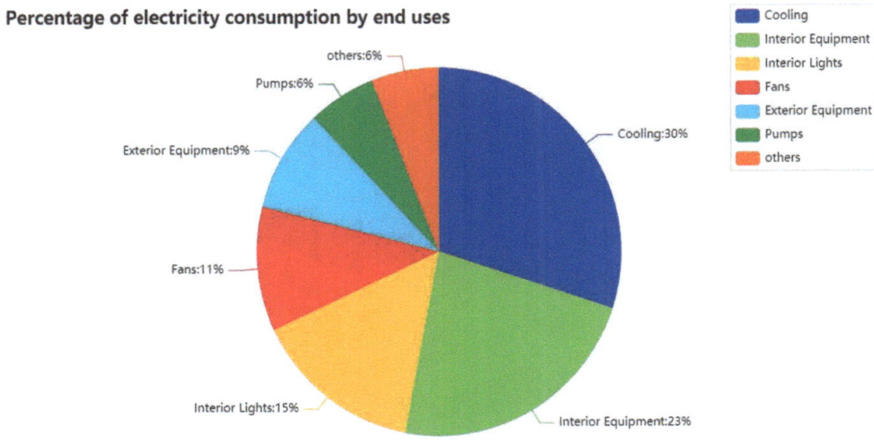

Fig. 2 Percentage distributions of electricity consumption by end uses for the hospital building

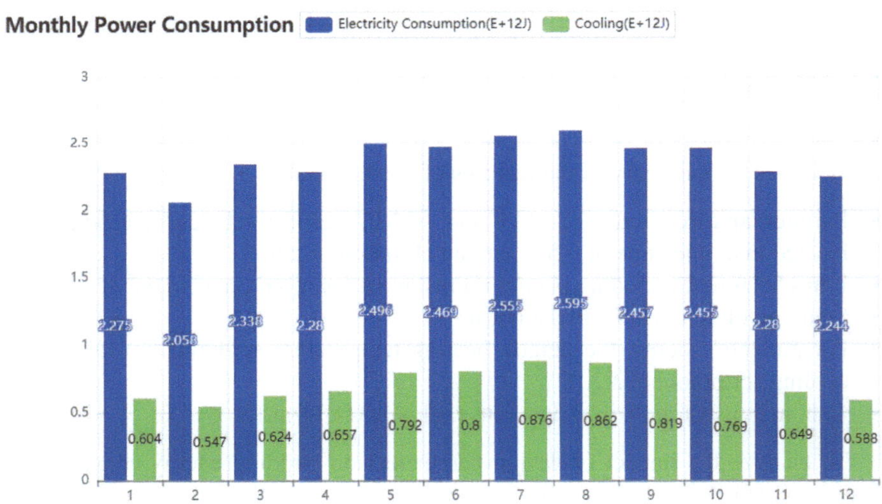

Fig. 3 Monthly total energy consumption and cooling electricity of the hospital building

others, the consumption of electricity by Cooling accounts for nearly one-third of the total energy consumption. This is followed by Interior Equipment and Interior Lights. This makes sense given that the hospital building is located in Florida, which has a hot climate throughout the year, with summers being particularly intense. As a result, the building consumes more cooling energy to maintain comfortable temperatures in patient rooms.

To investigate more granular level of cooling energy distribution, Fig. 3 shows monthly cooling electricity. As a result, a bar chart shows the difference of the total energy consumption of each month and that of cooling electricity, respectively. From

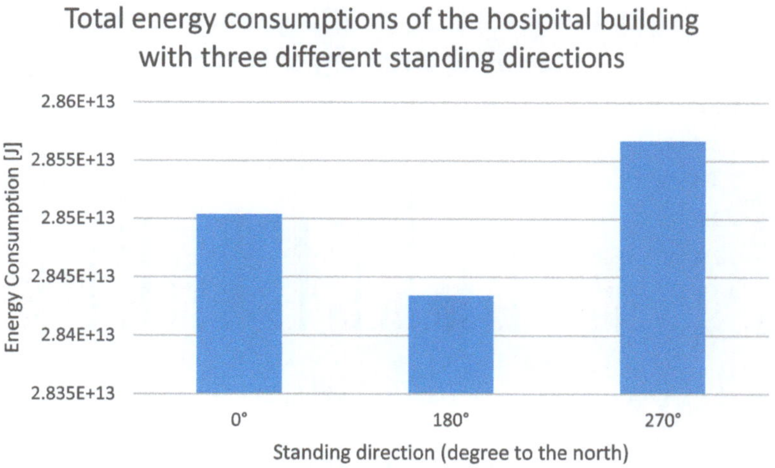

Fig. 4 Total energy consumptions of the hospital building with three different standing directions

Fig. 3, we can see that the cooling energy is the highest in July and August. It makes sense since they are the typical hot summer months in Florida, while the cooling energy reaches to the lowest point in February.

The following section illustrates the energy consumption differences when changing the design factors from three perspectives, standing direction, temperature setpoint, and construction material conductivity.

3.2 Standing Direction

With the same building design configurations, three different standing directions are simulated in the EnergyPlus to see the building energy consumption differences. As Fig. 4 shows, the standing direction of 180° consumes the least energy than other two options.

Figures 5 and 6 show the monthly total energy consumption of the hospital building with three different standing directions. Below are two main observations: (1) almost for every month, the building with 180° standing direction consumes the least energy than the other two. (2) Among all the months, February is odd with the least energy consumption comparing with the other months. Since cooling energy is the main contributing factor of the total energy consumption, the total energy and the cooling energy are both the least in February. However, when comparing the average temperature of January, February, and March, there is no observed obvious abnormal trend; in order, January is the coldest, with February being second and March the hottest. Therefore, the outside temperature is not the cause of the low energy consumption of February. The other possible cause can be changed staffing schedule and activities in the building in February comparing to other months.

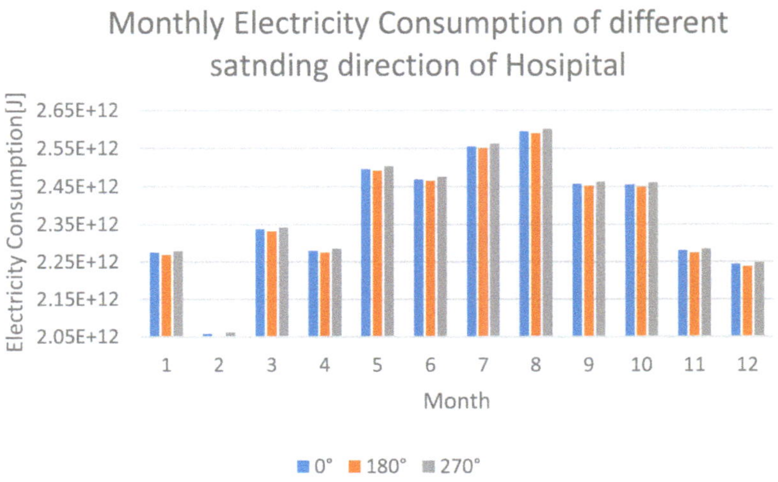

Fig. 5 Monthly total energy consumption for the hospital building with three different standing directions

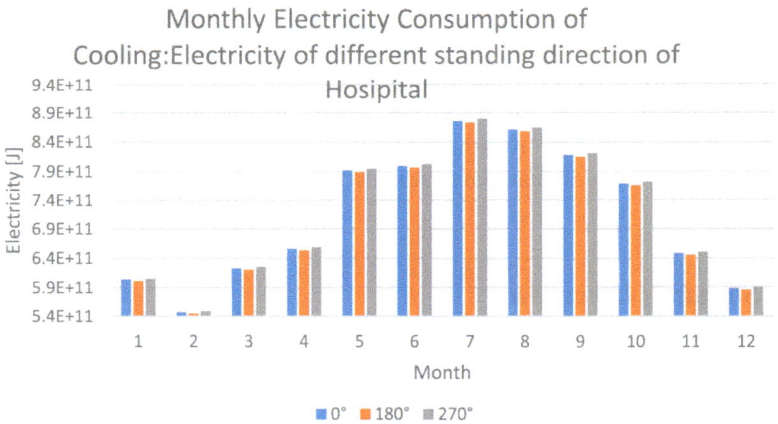

Fig. 6 Monthly cooling electricity for the hospital building with three different standing directions

In conclusion, from the energy saving perspective, the standing direction of 180° is the best choice among the three provided options. More simulations can be conducted to find the optimal standing direction that consumes the least energy consumption. The comparison of the energy consumption with these three different chosen standing directions illustrated in this section provides a quantitative guideline for users to see how the standing direction can influence the building energy consumption.

4 Recommendations

Using the standing direction as an example, through simulations, we have demonstrated the impact of changing design parameters during the early design phase on reducing a building's carbon footprint and improving its life cycle. We have also showcased the capabilities and utility of Energyplus, a building energy simulation software. For buildings in hot regions such as Miami, minimizing overall energy efficiency is crucial. By considering the standing direction of a building, natural ventilation and lighting can be utilized to reduce the need for mechanical cooling and lighting systems, and influence air and heat flow within the building, thereby improving overall energy efficiency. Also, we can strike a balance between energy consumption and human comfort condition like adjusting the temperature set-point basing on the psychrometric chart research. About human comfort conditions, there is a recommended indoor operation temperature setpoint, 22–27 °C [10]. Similarly, deep learning algorithms and artificial intelligence can also be used to design buildings in order to improve the whole life cycle of buildings [11–15]. In conclusion, this study provides valuable insights into the energy consumption of hospitals and outpatient buildings worldwide. Architects and builders can help reduce energy consumption by considering factors such as standing direction, temperature set-points, and insulation, promoting sustainability for these types of buildings.

5 Conclusions

This research project used EnergyPlus to investigate the energy consumption of healthcare buildings in the Miami area. The study found that building orientation significantly affects energy consumption. Adjusting it can reduce overall energy consumption. EnergyPlus is a widely used software tool for modeling and analyzing building energy consumption, which provides a deep understanding of how different factors affect the energy performance of buildings. However, this study has limitations as it only focuses on healthcare buildings in the Miami area and uses historical weather data to simulate the local climate. Including other types of healthcare buildings and using updated weather data in such a study can provide more accurate representations. Further research is needed on the effectiveness of the proposed strategy in reducing energy consumption and its impact on overall energy efficiency. Overall, this study provides valuable insights into reducing energy consumption in healthcare buildings and identifies potential strategies for architects and engineers to design energy-efficient and sustainable buildings.

References

1. U.S. Energy Information Administration. (2018). *2018 Commercial Buildings Energy Consumption Survey consumption and expenditures preliminary results.* https://www.eia.gov/consumption/commercial/. Last accessed August 12, 2022.
2. Liu, P., & Liu, Y. M. (2019). *Study on the factors affecting energy consumption in hospitals.* https://m.book118.com/html/2019/0626/8142034014002032.shtm?from=mip. Last accessed March 23, 2023.
3. Wang, W., Zmeureanu, R., & Rivard, H. (2005). Applying multi-objective genetic algorithm sin green building design optimization. *Building and Environment, 40*(11), 1512–1525.
4. Li, X., Zhou, Y., Sha, Y., Jia, G., Li, H., & Li, W. (2019). Urban heat Island impacts on building energy consumption: A review of approaches and findings. *Energy, 174*, 407–419. ISSN0360-5442. https://www.sciencedirect.com/science/article/pii/S0360544219303895. Last accessed September 19, 2023.
5. Wei, W., & Skye, H. M. (2021). Residential net-zero energy buildings: Review and perspective. *Renewable and Sustainable Energy Reviews, 142*, 110859. ISSN 1364-0321. https://www.sciencedirect.com/science/article/pii/S1364032121001532. Last accessed September 19, 2023.
6. Crawleya, D. B., Lawrieb, L. K., Winkelmannc, F. C., Buhlc, W. F., Huangc, Y. J., Pedersend, C. O., Strandd, R. K., Liesend, R. J., Fishere, D. E., Wittef, M. J., & Glazerf, J. (2001). EnergyPlus: Creating a new-generation building energy simulation program. *Energy and Buildings, 33*(4), 319–331.
7. Yu, N., & Paolucci, S. (2017). Model-based design optimization and predictive control to minimize energy consumption of a building[C]. In *Proceedings of CHT-17 ICHMT international symposium on advances in computational heat transfer*. Begel House Inc..
8. Dombayci, Ö. A. (2007). The environmental impact of optimum insulation thickness for external walls of buildings. *Building and Environment, 42*(11), 3855–3859.
9. Bui, D.-K., Nguyen, T. N., Ghazlan, A., Ngo, N.-T., & Ngo, T. D. (2020). Enhancing building energy efficiency by adaptive façade: A computational optimization approach. *Applied Energy, 265*, 114797. ISSN 0306-2619. https://www.sciencedirect.com/science/article/pii/S0306261920303093
10. Zhou, N., Khanna, N., Feng, W., Hong, L. X., Fridley, D., Creyts, J., Franconi, E., Torbert, R., & Ke, Y. (2014). Cost-effective options for transforming the Chinese building sector. In *2014 ACEEE summer study on energy efficiency in buildings* (Vol. 3, pp. 367–377). American Council for an Energy-Efficient Economy.
11. Wellcare Global. (2021). *What is the human comfort zone for temperature and humidity?*https://blog.wellcare-global.com/blog/what-is-the-human-comfort-zone-for-temperature-and-humidity. Last accessed April 4, 2023.
12. Olu-Ajayi, R., Alaka, H., Sulaimon, I., et al. (2022). Building energy consumption prediction for residential buildings using deep learning and other machine learning techniques[J]. *Journal of Building Engineering, 45*, 103406.
13. Himeur, Y., Ghanem, K., Alsalemi, A., et al. (2021). Artificial intelligence based anomaly detection of energy consumption in buildings: A review, current trends and new perspectives[J]. *Applied Energy, 287*, 116601.
14. Farzaneh, H., Malehmirchegini, L., Bejan, A., et al. (2021). Artificial intelligence evolution in smart buildings for energy efficiency[J]. *Applied Sciences, 11*(2), 763.
15. Alzoubi, A. (2022). Machine learning for intelligent energy consumption in smart homes[J]. *International Journal of Computations, Information and Manufacturing (IJCIM), 2*(1). https://doi.org/10.54489/ijcim.v2i1.75

Evaluation and Identification of Waste Heat Utilization Pathways: A Review

Jan-Niklas Gerdes and Alexander Sauer

1 Introduction

The industrial sector can significantly contribute to the reduction of greenhouse gases since it accounts for about one-third of the global final energy demand [1] and about 475 TWh per year in Germany alone [2]. The challenge of rising energy costs due to the current energy crisis implies the need for manufacturing companies to improve their energy efficiency to remain competitive and comply with future environmental regulations [3]. In Germany, about 70% of all energy consumption in the industrial sector can be attributed to the thermal energy demand [4] and is typically provided by fossil fuels like coal and natural gas [5]. However, substantial waste heat potential in the industrial sector remains unused [6]. In Germany alone, the waste heat potential is estimated to be at least 140 TWh per year, with around 132 TWh considered useful waste heat at a temperature above 60 °C [7]. Depending on the temperature level of the available waste heat, this potential can be used on-site through the connection of thermal energy streams or by conversion technologies, like heat pumps or absorption chillers [8]. Due to this complexity, methodological approaches to the identification of optimal waste heat utilization have been developed, such as the pinch analysis, which is widely used to design chemical plants [9]. In recent years, mathematical and simulative approaches have been developed to optimize waste heat utilization according to specific key performance indexes (KPIs) [8]. These approaches aim to model the processes within the plant and calculate the optimal implementation of waste heat utilization [9], but differ widely in the scope of technology and timeframe considered.

J.-N. Gerdes (✉) · A. Sauer
University Stuttgart Institute for Energy Efficiency in Production, Stuttgart, Germany

Fraunhofer Institute for Manufacturing Engineering and Automation, Stuttgart, Germany
e-mail: jan-niklas.gerdes@ipa.fraunhofer.de

P. Pong (ed.), *Renewable Energy Resources and Conservation*, Green Energy and Technology, https://doi.org/10.1007/978-3-031-59005-4_23

Therefore, this chapter aims to identify trends in methodological approaches for waste heat utilization and indicate research gaps. A review of the academic literature regarding the identification of optimal waste heat utilization pathways in industrial energy systems was conducted.

2 Industrial Waste Heat Utilization Pathways

Utilization options for industrial waste heat are numerous and generally dependent on the available temperature level [8]. Furthermore, additional situational factors, such as the plant energy demand and technical characteristics of the waste heat source and sink, affect the choice of optimal utilization pathway. Thus, methodological design approaches are necessary to determine optimal waste heat recovery.

2.1 Waste Heat Utilization Technologies

Recovered waste heat potential can be harnessed directly through heat exchangers or converted into other forms of energy, like electricity and chilling through conversion technologies [9]. Furthermore, the available thermal energy can be used either on-site or distributed off-site.

Waste heat can be converted into electrical power through multiple conversion technologies, with organic Rankine cycle (ORC)-based plants being the most prominent application. This is due to their flexibility, simplicity, and relatively low temperature requirements [10]. These plants can work with heat sources at relatively low temperatures by using organic working fluids with vaporization temperatures far lower than that of water [9]. The generated electricity can then be used to reduce the electricity demand of the entire factory and the emissions associated with the imported electricity. However, ORC plants are significantly more expensive than conventional steam power plants due to lower efficiency and higher equipment requirements [11].

Chilling power can be supplied to the factory using waste heat-driven chillers, such as adsorption or absorption chillers [12]. As conventional chillers use compression-based cycles to produce chilling power through electricity, waste heat-driven technologies can significantly reduce energy costs and emissions. This is due to the relatively high emissions and costs associated with electricity imports compared to waste heat. When coupled with a waste heat source or combined heat and power (CHP) plant, these chillers enable economically feasible chilling power generation, despite higher investment [13].

Waste heat can be recovered directly into usable thermal energy with heat exchangers, stored in thermal energy storage units [14], or by upgrading it through heat pumps, commonly driven by electric power. While heat pumps can operate very efficiently, they cannot reach higher temperature levels and commonly operates at temperatures up to 110 °C [9]. In addition to the efficiency benefits, implementing heat pumps in the industrial energy system allows the usage of the inherent energy flexibility potential [15]. If the recovered thermal energy cannot be used in the site's energy system, it can be exported off-site and used either in distribution networks in the industrial site [16] or in district heating networks (DHS) for communal heating [17]. As thermal energy and heating are usually supplied by burning fossil fuels, this creates significant emission-saving potential for the industrial sector.

2.2 Design Approaches

Due to the relatively high amount of waste heat utilization pathways and the complexity of integration, multiple design approaches for optimizing waste heat utilization have been developed. One of the most prominent design approaches is the pinch analysis, which enables the identification of the optimal recovery of waste heat in continuous processes through the use of composite curves of hot and cool streams [18]. This approach has been further developed to enable the optimization of waste heat utilization in batch processes but is limited by the complexity of time-discrete variations [9]. Therefore, optimization-based approaches have been proposed to determine the optimum mathematically. One such approach is the OMNIUM method, which solves an optimization problem using the Hungarian method [19] and aims to determine the economically optimal solution [9]. Based on this method, the KOARiiS approach was developed to determine a more practical optimum for waste heat integration and reduce barriers to utilization [20]. Generally, these methodologies investigate the optimal interconnectivity of heat flows, with only a few approaches examining energy conversion applications. One such approach was developed by Oluleye, determining the optimal waste heat utilization in the process industry by considering conversion technologies and specific key performance indicators (KPIs) [21]. While multiple forms of useable energy were included, no time-discrete data was used.

3 Methodology

In this chapter, the framework of the scientific literature review is presented. To classify the works analyzed and identify relevant research gaps regarding approaches for waste heat utilization identification, a brief methodology is illustrated., The scientific literature is reviewed according to the investigated waste heat utilization pathways, the temporal resolution of the data used, and the assessment dimensions, like economic, ecologic, or technical KPIs used for evaluation.

The assessment criteria for this literature review are shown in Table 1. When evaluating the waste heat utilization pathways, the literature can be divided into three categories, depending on the number of energy forms investigated with the

Table 1 Assessment criteria for waste heat utilization approaches

Assessment criteria			
	Single	Multiple	All
Waste heat utilization pathways	Only thermal utilization	Thermal utilization and conversion into electricity	Thermal utilization and conversion into all energy forms
	Continuous	Partial	Complete
Temporal resolution of data used	Energy data is assumed to be continuous	Energy data is assumed to be continuous, but attempts to take daily or seasonal variations into account were made	Energy data is time discrete and variations on daily and seasonal basis recorded
	Single	Multiple	All
Evaluation dimensions	Only one dimension was used	Multiple dimensions were used	All dimensions were used

approach. Generally, many approaches optimize waste heat recovery as useful thermal energy and disregard conversion into other forms of energy, while some approaches further investigate the conversion into only electricity. Therefore, the more pathways and energy forms are investigated, the higher the score. Another criterion used for this chapter is the temporal resolution of the data used in the optimization approaches. This can vary from continuous energy data to time-discrete measurement data from sensors and data collectors. As more time-discrete data considers the actual daily and seasonal variations, the results regarding waste heat recovery will be more reliable and reasonable. With the necessity for simultaneousness between the heat sink and source in waste heat utilization concepts, this criterion is very important. A further distinction between design approaches is based on the evaluation dimensions that consider economic, ecologic (i.e., emission saving potential), and technical (i.e., efficiency) aspects.

4 Results

Using the assessment criteria introduced in the previous chapter, the scientific literature reviewed can be categorized into nine distinct sections depending on the temporal resolution of the data used for waste heat utilization and the number of waste heat utilization pathways explored. Further separation was made on the number of evaluation dimensions with the resulting matrix presented in Fig. 1. In Fig. 1(I), a significant amount of literature is focused on the implementation and optimization of waste heat recovery (WHR) in the form of district heating (DH) [22–31] or by the distribution between plants in an industrial site [32]. As the distribution of thermal energy generally requires constant thermal energy

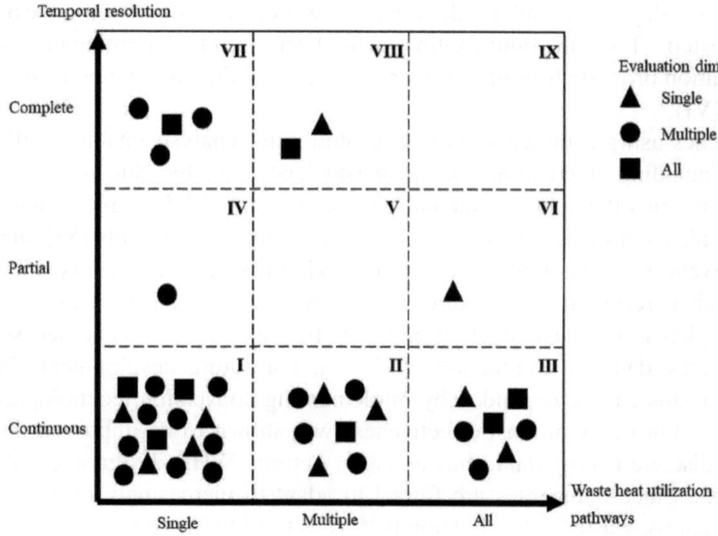

Fig. 1 Evaluation matrix divided into nine different sections

flow, the methodologies of papers in this section used continuous data for their calculations. Other literature in this section analyzed the implementation of WHR as an energy efficiency measure for industries, therefore using constant energy data and analyzing only recuperation in the form of heating energy [33–35]. This approach is expanded in Fig. 1(II), taking into account the option of WHR into electricity through ORC plants. However, the literature categorized in this section generally deals with implementing ORC into specific use cases or industries and their potential applications, thus using continuous data for evaluation [36–43]. Figure 1(III) explores all possible waste heat utilization pathways with differing methodologies proposed in the scientific literature. Due to the differences in temperature ranges for conversion technologies, the temperature level of the waste heat available must be considered, resulting in distinct categories of temperature level and ranking of technologies [8, 44, 45]. This aspect further influences design methodologies proposed in [46–50], investigating the optimal WHR design depending on the waste heat's temperature level and the production plant's energy demand. Generally, methodologies in this section use an approach similar to the pinch analysis for examination, enabling cascade solutions but limited to continuous energy data for energy flows.

Zhang et al. [51] propose an optimization method for the WHR implementation in industrial parks through waste heat transportation systems, combining the possibility of utilization in other companies and in DH networks. To assess discontinuous waste heat implementation, the data was separated into time slices, thus being categorized into Fig. 1(IV). Similarly, Wang et al. [52] propose an optimization method using energy data for 1 entire year, approximated by using

2 characteristic days per month for the week and weekends, thus reducing the dataset to 576 timesteps. Then, the optimization method was used to determine the optimal implementation of waste heat recovery and conversion, therefore being categorized into Fig. 1(VI).

Approaches using complete temporal resolution for analysis enable WHR optimization, including fluctuations in the waste heat available due to production characteristics or other plant behavior. This can assist WHR implementation in production plants through improvements to the decision support tools [53], increase cost-effectiveness of heat pump configuration [54] and ORC plant composition [55, 56]. Another increasingly important examination aspect is the energy flexibility of production plants, which can be assisted by WHR implementation and necessitates discrete energy data for examination [57, 58]. For future developments in this subject area, this can be expanded by implementing conversion technologies into this analysis. Furthermore, cost-effectiveness was shown to be improved through the use of discrete energy data, thus assisting optimal WHR design and reducing energy costs. Since companies are forced to adapt to increasingly more difficult market situations, waste heat utilization pathways need to be assessed according to multiple evaluation dimensions to ensure the communication of benefits and ease of implementation.

5 Conclusion

This chapter reviews the scientific literature on methodologies regarding waste heat utilization in industrial companies, categorizing it by a temporal resolution of energy data analyzed and the number of waste heat utilization pathways. While plenty of methodologies evaluate the implementation of WHR through either recovery or conversion technologies, it was shown that most use continuous data. This requires little fluctuation in the waste heat available and is thus generally applicable to process industries. When using energy data with the complete temporal resolution, the implementation of WHR was typically more cost-effective than the evaluation using continuous data. Furthermore, the investigation of energy flexibility aspects and their influence on the cost-effectiveness of the energy system is possible by discrete energy data, granting higher cost saving potential. With increasing volatility of energy production and supply difficulties, especially of thermal energy, industrial companies will need to adopt energy efficiency measures like WHR in the near future. However, due to the volatility of waste heat and available conversion technologies, complexity hinders the implementation. To reduce this complexity and assist decision-making, a methodology to evaluate waste heat utilization pathways on the basis of characteristic plant behavior and energy demand will be necessary, requiring the ability to assess all potential conversion technologies, as well as multiple evaluation dimensions.

Acknowledgments The authors gratefully acknowledge the financial support of the German Federal Ministry of Economic Affairs and Climate Action (BMWK), the project supervision of the Project Management Jülich (PtJ) for the project "FlexGUIde," and the companies that provided their data for the purpose of this research.

References

1. REN21. (2022). *Renewables 2022 Global Status Report.*
2. Landesamt für Natur, Umwelt und Verbraucherschutz. (2019). Potenzialstudie Industrielle Abwärme.
3. Bauer, D., Hieronymus, A., Kaymakci, C., et al. (2021). Wie IT die Energieflexibilitätsvermarktung von Industrieunternehmen ermöglicht und die Energiewende unterstützt. *HMD, 58*, 102–115. https://doi.org/10.1365/s40702-020-00679-8
4. (2012). *Energie in Zahlen: Arbeit und Leistungen der AG Energiebilanzen* (1. Aufl). AG Energiebilanzen e. V.
5. Bundesamt, U. (2022). *Energieverbrauch für fossile und erneuerbare Wärme*. https://www.umweltbundesamt.de/daten/energie/energieverbrauch-fuer-fossile-erneuerbare-waerme#warmeverbrauch-und-erzeugung-nach-sektoren. Accessed May 9, 2023.
6. Pehnt, M., Bödeker, J., Arens, M., et al. (2011). *Industrial Waste Heat – Tapping into a neglected efficiency potential.* Fraunhofer-Gesellschaft.
7. Brueckner, S., Miró, L., Cabeza, L. F., et al. (2014). Methods to estimate the industrial waste heat potential of regions – A categorization and literature review. *Renewable and Sustainable Energy Reviews, 38*, 164–171. https://doi.org/10.1016/j.rser.2014.04.078
8. Oluleye, G., Jiang, N., Smith, R., et al. (2017). A novel screening framework for waste heat utilization technologies. *Energy, 125*, 367–381. https://doi.org/10.1016/j.energy.2017.02.119
9. Radgen, P., Hufendiek, K., & Blesl, M. (Eds.). (2022). *Abwärmepotentiale in der Industrie: Konzepte zur Nutzung im Mittel- und Niedrigtemperaturbereich* (1. Auflage). Beuth Praxis. Beuth Verlag.
10. Gao, P., Jiang, L., Wang, L. W., et al. (2015). Simulation and experiments on an ORC system with different scroll expanders based on energy and exergy analysis. *Applied Thermal Engineering, 75*, 880–888. https://doi.org/10.1016/j.applthermaleng.2014.10.044
11. Mahmoudi, A., Fazli, M., & Morad, M. R. (2018). A recent review of waste heat recovery by Organic Rankine Cycle. *Applied Thermal Engineering, 143*, 660–675. https://doi.org/10.1016/j.applthermaleng.2018.07.136
12. Herold, K. E., Radermacher, R., & Klein, S. A. (2016). *Absorption chillers and heat pumps* (2nd ed.). CRC Press.
13. Dehli, M. (2020). *Energieeffizienz in Industrie, Dienstleistung und Gewerbe: Energietechnische Optimierungskonzepte für Unternehmen*. Springer Vieweg.
14. Emde, A. (2023). *Techno-ökonomische Bewertung von energieträgerübergreifenden hybriden Energiespeichern*. Universität Stuttgart.
15. Sadjjadi, B. S., Gerdes, J.-N., & Sauer, A. (2023). Energy flexible heat pumps in industrial energy systems: A review. *Energy Reports, 9*, 386–394. https://doi.org/10.1016/j.egyr.2022.12.110
16. Gerdes, J.-N., Munder, M., & Sauer, A. (2023). Evaluation of technical and economic potential of waste heat distribution networks in industrial sites. *Energy Reports, 9*, 219–226. https://doi.org/10.1016/j.egyr.2022.12.112
17. Li, H., Sun, Q., Zhang, Q., et al. (2015). A review of the pricing mechanisms for district heating systems. *Renewable and Sustainable Energy Reviews, 42*, 56–65. https://doi.org/10.1016/j.rser.2014.10.003

18. Klemeš, J. J., & Kravanja, Z. (2013). Forty years of heat integration: Pinch analysis (PA) and mathematical programming (MP). *Current Opinion in Chemical Engineering, 2*, 461–474. https://doi.org/10.1016/j.coche.2013.10.003
19. Hellwig, T. (1998). *OMNIUM: ein Verfahren zur Optimierung der Abwärmenutzung in Industriebetrieben* (Dissertation). Universität Stuttgart.
20. Heyden, E. (2016). *Kostenoptimale Abwärmerückgewinnung durch integriert-iteratives Systemdesign (KOARiiS): ein Verfahren zur energetisch-ökonomischen Bewertung industrieller Abwärmepotenziale*. Universität Stuttgart.
21. Oluleye, O. (2016). *Integration of waste heat recovery in process sites* (Dissertation). The University of Manchester.
22. Torío, H., & Schmidt, D. (2010). Development of system concepts for improving the performance of a waste heat district heating network with exergy analysis. *Energy and Buildings, 42*, 1601–1609. https://doi.org/10.1016/j.enbuild.2010.04.002
23. Fitó, J., Hodencq, S., Ramousse, J., et al. (2020). Energy- and exergy-based optimal designs of a low-temperature industrial waste heat recovery system in district heating. *Energy Conversion and Management, 211*, 112753. https://doi.org/10.1016/j.enconman.2020.112753
24. Fang, H., Xia, J., & Jiang, Y. (2015). Key issues and solutions in a district heating system using low-grade industrial waste heat. *Energy, 86*, 589–602. https://doi.org/10.1016/j.energy.2015.04.052
25. Moser, S., Puschnigg, S., & Rodin, V. (2020). Designing the Heat Merit Order to determine the value of industrial waste heat for district heating systems. *Energy, 200*, 117579. https://doi.org/10.1016/j.energy.2020.117579
26. Fang, H., Xia, J., Zhu, K., et al. (2013). Industrial waste heat utilization for low temperature district heating. *Energy Policy, 62*, 236–246. https://doi.org/10.1016/j.enpol.2013.06.104
27. Pelda, J., Stelter, F., & Holler, S. (2020). Potential of integrating industrial waste heat and solar thermal energy into district heating networks in Germany. *Energy, 203*, 117812. https://doi.org/10.1016/j.energy.2020.117812
28. Kim, H.-W., Dong, L., Choi, A. E. S., et al. (2018). Co-benefit potential of industrial and urban symbiosis using waste heat from industrial park in Ulsan, Korea. *Resources, Conservation and Recycling, 135*, 225–234. https://doi.org/10.1016/j.resconrec.2017.09.027
29. Du, K., Calautit, J., Eames, P., et al. (2021). A state-of-the-art review of the application of phase change materials (PCM) in Mobilized-Thermal Energy Storage (M-TES) for recovering low-temperature industrial waste heat (IWH) for distributed heat supply. *Renewable Energy, 168*, 1040–1057. https://doi.org/10.1016/j.renene.2020.12.057
30. Wahlroos, M., Pärssinen, M., Rinne, S., et al. (2018). Future views on waste heat utilization – Case of data centers in Northern Europe. *Renewable and Sustainable Energy Reviews, 82*, 1749–1764. https://doi.org/10.1016/j.rser.2017.10.058
31. Wang, J., Wang, Z., Zhou, D., et al. (2019). Key issues and novel optimization approaches of industrial waste heat recovery in district heating systems. *Energy, 188*, 116005. https://doi.org/10.1016/j.energy.2019.116005
32. Chae, S. H., Kim, S. H., Yoon, S.-G., et al. (2010). Optimization of a waste heat utilization network in an eco-industrial park. *Applied Energy, 87*, 1978–1988. https://doi.org/10.1016/j.apenergy.2009.12.003
33. Oluleye, G., Smith, R., & Jobson, M. (2016). Modelling and screening heat pump options for the exploitation of low grade waste heat in process sites. *Applied Energy, 169*, 267–286. https://doi.org/10.1016/j.apenergy.2016.02.015
34. Nowicki, C., & Gosselin, L. (2012). An overview of opportunities for waste heat recovery and thermal integration in the primary aluminum industry. *JOM, 64*, 990–996. https://doi.org/10.1007/s11837-012-0367-4
35. Kurle, D., Schulze, C., Herrmann, C., et al. (2016). Unlocking waste heat potentials in manufacturing. *Procedia CIRP, 48*, 289–294. https://doi.org/10.1016/j.procir.2016.03.107
36. Agathokleous, R., Bianchi, G., Panayiotou, G., et al. (2019). Waste Heat Recovery in the EU industry and proposed new technologies. *Energy Procedia, 161*, 489–496. https://doi.org/10.1016/j.egypro.2019.02.064

37. Jin, Y., Gao, N., & Zhu, T. (2019). Techno-economic analysis on a new conceptual design of waste heat recovery for boiler exhaust flue gas of coal-fired power plants. *Energy Conversion and Management, 200*, 112097. https://doi.org/10.1016/j.enconman.2019.112097
38. Jouhara, H., Khordehgah, N., Almahmoud, S., et al. (2018). Waste heat recovery technologies and applications. *Thermal Science and Engineering Progress, 6*, 268–289. https://doi.org/10.1016/j.tsep.2018.04.017
39. Loni, R., Najafi, G., Bellos, E., et al. (2021). A review of industrial waste heat recovery system for power generation with Organic Rankine Cycle: Recent challenges and future outlook. *Journal of Cleaner Production, 287*, 125070. https://doi.org/10.1016/j.jclepro.2020.125070
40. Lu, H., Price, L., & Zhang, Q. (2016). Capturing the invisible resource: Analysis of waste heat potential in Chinese industry. *Applied Energy, 161*, 497–511. https://doi.org/10.1016/j.apenergy.2015.10.060
41. Sani, M. M., Noorpoor, A., & Motlagh, M. S. (2020). Multi objective optimization of waste heat recovery in cement industry (a case study). *Journal of Thermal Engineering, 6*, 604–618. https://doi.org/10.18186/thermal.764536
42. Shi, S., Wang, Y., Wang, Y., et al. (2022). A new optimization method for cooling systems considering low-temperature waste heat utilization in a polysilicon industry. *Energy, 238*, 121800. https://doi.org/10.1016/j.energy.2021.121800
43. Shu, G., Yu, G., Tian, H., et al. (2014). A Multi-Approach Evaluation System (MA-ES) of Organic Rankine Cycles (ORC) used in waste heat utilization. *Applied Energy, 132*, 325–338. https://doi.org/10.1016/j.apenergy.2014.07.007
44. Su, Z., Zhang, M., Xu, P., et al. (2021). Opportunities and strategies for multigrade waste heat utilization in various industries: A recent review. *Energy Conversion and Management, 229*, 113769. https://doi.org/10.1016/j.enconman.2020.113769
45. Xu, Z. Y., Wang, R. Z., & Yang, C. (2019). Perspectives for low-temperature waste heat recovery. *Energy, 176*, 1037–1043. https://doi.org/10.1016/j.energy.2019.04.001
46. Oluleye, G., & Jobson, M. (2015). Optimisation-based design of site waste heat recovery systems. In *ECOS 2015 – 28th international conference on efficiency, cost, optimization, simulation and environmental impact of energy systems*.
47. Oluleye, G., Jobson, M., & Smith, R. (2015). A hierarchical approach for evaluating and selecting waste heat utilization opportunities. *Energy, 90*, 5–23. https://doi.org/10.1016/j.energy.2015.05.086
48. Oluleye, G., Jobson, M., & Smith, R. (2016). Process integration of waste heat upgrading technologies. *Process Safety and Environmental Protection, 103*, 315–333. https://doi.org/10.1016/j.psep.2016.02.003
49. Oluleye, G., Jobson, M., Smith, R., et al. (2016). Evaluating the potential of process sites for waste heat recovery. *Applied Energy, 161*, 627–646. https://doi.org/10.1016/j.apenergy.2015.07.011
50. Giordano, L., & Benedetti, M. (2022). A methodology for the identification and characterization of low-temperature waste heat sources and sinks in industrial processes: Application in the Italian dairy sector. *Energies, 15*, 155. https://doi.org/10.3390/en15010155
51. Zhang, C., Zhou, L., Chhabra, P., et al. (2016). A novel methodology for the design of waste heat recovery network in eco-industrial park using techno-economic analysis and multi-objective optimization. *Applied Energy, 184*, 88–102. https://doi.org/10.1016/j.apenergy.2016.10.016
52. Wang, X., Jin, M., Feng, W., et al. (2018). Cascade energy optimization for waste heat recovery in distributed energy systems. *Applied Energy, 230*, 679–695. https://doi.org/10.1016/j.apenergy.2018.08.124
53. Simeone, A., Luo, Y., Woolley, E., et al. (2016). A decision support system for waste heat recovery in manufacturing. *CIRP Annals, 65*, 21–24. https://doi.org/10.1016/j.cirp.2016.04.034

54. Kosmadakis, G., Arpagaus, C., Neofytou, P., et al. (2020). Techno-economic analysis of high-temperature heat pumps with low-global warming potential refrigerants for upgrading waste heat up to 150 °C. *Energy Conversion and Management, 226*, 113488. https://doi.org/10.1016/j.enconman.2020.113488
55. Mezzera, F., Fattori, F., Dénarié, A., et al. (2021). Waste-heat utilization potential in a hydrogen-based energy system – An exploratory focus on Italy. *International Journal of Sustainable Energy Planning and Management, 31*, 95–108. https://doi.org/10.5278/ijsepm.6292
56. Pili, R., Romagnoli, A., Spliethoff, H., et al. (2017). Techno-economic analysis of waste heat recovery with ORC from fluctuating industrial sources. *Energy Procedia, 129*, 503–510. https://doi.org/10.1016/j.egypro.2017.09.170
57. Li, D., Wang, J., Ding, Y., et al. (2019). Dynamic thermal management for industrial waste heat recovery based on phase change material thermal storage. *Applied Energy, 236*, 1168–1182. https://doi.org/10.1016/j.apenergy.2018.12.040
58. Köfinger, M., Schmidt, R. R., Basciotti, D., et al. (2018). Simulation based evaluation of large scale waste heat utilization in urban district heating networks: Optimized integration and operation of a seasonal storage. *Energy, 159*, 1161–1174. https://doi.org/10.1016/j.energy.2018.06.192

Evaluation of a Heat Pump Integration in the District Heating Supply of a Production Facility

Bijan Sadjjadi-Ortlieb and Alexander Sauer

1 Introduction

The EU Commission has established ambitious goals for reducing carbon emissions in industrial operations as part of the Green Deal. Their objective is to achieve climate neutrality by 2050. Achieving this goal entails increasing the proportion of renewable energy in the electricity grid by 40% by 2030 [1], as well as electrifying industrial processes in the manufacturing sector by up to 50% [2]. Industrial heat pumps, utilizing electrical energy, can generate thermal energy for industrial processes [3] and achieve temperatures of up to 150 °C [4]. This work aims to investigate the technical integration of an industrial heat pump into the heating system of a production hall. Specifically, it will analyse the energy-related differences compared to the reference case, where thermal energy is exclusively supplied through a district heating system.

2 Fundamentals

This work primarily focuses on the integration of a closed-system electric compression heat pump, which is widely employed in various industrial applications [5]. Industrial heat pumps play a crucial role in providing thermal energy, either heat or cold, for production processes. The energy input, in the form of electrical energy ($P_{electric}$), is used to compress the refrigerant by a compressor. The transfer of

B. Sadjjadi-Ortlieb (✉) · A. Sauer
Institute for Energy Efficiency in Production, University of Stuttgart, Stuttgart, Germany

Fraunhofer Institute for Manufacturing Engineering and Automation, Stuttgart, Germany
e-mail: bijan.viktor.sadjjadi-ortlieb@ipa.fraunhofer.de

P. Pong (ed.), *Renewable Energy Resources and Conservation*, Green Energy and Technology, https://doi.org/10.1007/978-3-031-59005-4_24

thermal energy for heating purposes ($\dot{Q}_{sink}$) occurs during the condensation process of the gaseous refrigerant. Subsequently, an expansion machine expands condensed refrigerant. Finally, the medium is evaporated in a heat exchanger utilizing waste or ambient heat ($\dot{Q}_{source}$), thereby completing the circuit of the basic vapour compression cycle. The heat pump process can be characterized by the coefficient of performance (COP). The basis for calculating the COP of a heat pump is described in Eq. (1) according to [5].

$$\mathrm{COP} = \frac{\dot{Q}_{sink}}{P_{electric}} = \eta \cdot \frac{T_{sink}}{T_{sink} - T_{source}} \tag{1}$$

The COP refers to the ratio between converted heating energy and the electrical energy consumed. Furthermore, the COP of compression heat pumps can be estimated using the efficiency factor (η), which usually ranges between 0.4 and 0.6 [5], the temperature at the evaporator (T_{source}) in the source circuit, and the temperature at the condenser (T_{sink}) in the sink circuit (see Eq. 1). A further important metric that assesses the heat pump's performance throughout the year is the annual COP (ACOP). It compares the heating energy produced over the year to the electrical energy consumed. Considering the annual thermal energy consumption, it can also display seasonal fluctuations. To compare the heat pump's integration to the reference system, an overall system efficiency (ω) is introduced. It is calculated according to [6] using Eq. (2), which relates the energy effort (Q_{effort}) to the benefit ($Q_{benefit}$). Q_{effort} describes the primary energy amount, which is supplied to the system and $Q_{benefit}$ describes the effective energy amount, which is needed in order to supply the heating process. The energy balance is established around the factory, excluding conversion losses in the district heating supply from consideration.

$$\omega = \frac{Q_{effort}}{Q_{benefit}} \tag{2}$$

3 Modelling of the Thermal System

This chapter describes the modelling of the energy system. The first section deals with modelling the existing thermal system. The next step describes the procedure for modelling the heat pump.

3.1 Heat Supply Network

The system used as a reference in this chapter pertains to a production hall where heat is supplied through a district heating (DH) pipeline. This reference

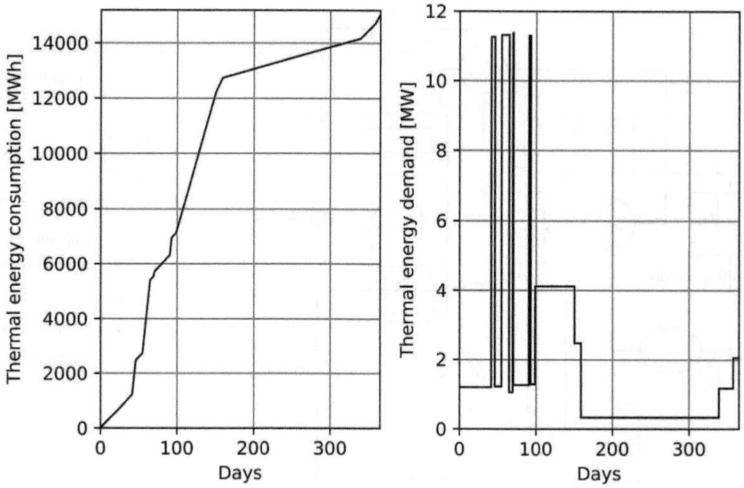

Fig. 1 Illustration of the annual thermal energy demand

system comprises a generation circuit, the district heating supply, and a thermal consumption network. A heat exchanger connects the consumption network to the district heating circuit.

Figure 1 illustrates the annual consumption profile of the heat supply. The DH delivers heat to the production hall at a flow temperature of 120 °C. A heat exchanger connects the district heating circuit. The district heating circuit is operated with a flow temperature control mechanism to maintain a constant temperature even when there are fluctuations in heat demand [4]. Additionally, it is assumed that the distribution network maintains a constant temperature, which is ensured by controlling the pumps in the consumer circuit using a mass flow control mechanism [5].

Further, the heat supply system is modelled using the object-oriented Modelica modelling language and the simulation programme Dymola. The heat supply system consists of multiple circuits independently driven by a pump and interconnected through heat exchangers. As depicted in Fig. 2, the consumption circuit receives actual consumption data through a generic consumer model.

Both the district heating circuit and the heat pump (HP) sink circuit can optionally supply the consumption circuit. The heat pump source circuit is connected downstream of the consumer, ensuring that the heat pump sink circuit is linked to the consumer circuit's return flow. The different circuits are controlled by a temperature controller that regulates the respective energy converters and pumps. Additionally, the heat pump, with a thermal output power of 1 MW, is activated when the return flow of the district heating circuit exceeds 60 °C.

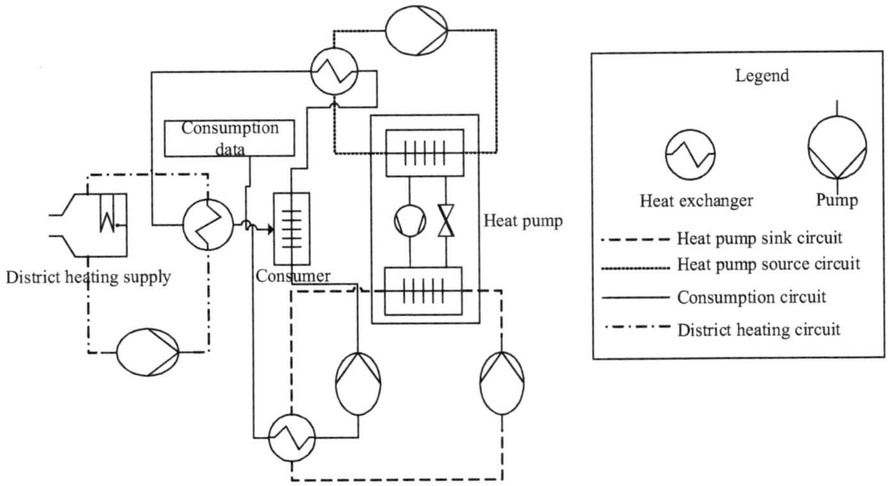

Fig. 2 Schematic illustration of the simulation model

3.2 Heat Pump

Regarding the heat pump, several methods can be used to model it. Semi-empirical COP models based on regression analysis, according to Jesper et al., Schlosser, and Wolf [7–9], can be used to model industrial heat pumps in addition to physical models. In this way, a dynamic system behaviour of the heat pump under fluctuating process boundary conditions can be represented, whereby the modelling focuses on the integration of the heat pump and represents the heat pump as a complete system. The COP of the heat pump has been modelled in this work using Eq. (3) as a function of the temperature lift ΔT_{lift}, the difference between the heat source and heat sink temperatures, and as a function of the sink temperature T_{sink}.

$$\mathrm{COP} = a \cdot (\Delta T_{\mathrm{lift}} + 2b)^{c} \cdot (T_{\mathrm{sink}} + b)^{d} \tag{3}$$

$$\dot{Q}_{\mathrm{sink}} = \dot{Q}_{\mathrm{source}} + P_{\mathrm{electric}} \tag{4}$$

The fitting parameters *a*, *b*, *c*, and *d* from the regression analysis, according to Schlosser et al. [8], and the validity ranges of the COP modelling for high temperature heat pumps are shown in Table 1.

The fitting parameters *a*, *b*, *c*, and *d* are obtained according to Schlosser et al. [8], by performing a regression analysis which links the COP from various market available heat pumps to the temperature lift and the temperature at the sink.

Table 1 Recommended fitting parameters for high-temperature heat pumps as a function of temperature

Heat pump type	Range of validity	Fitting parameter
High-temperature heat pump	$25\ ^\circ\text{C} \leq T_{\text{source}} \leq 110\ ^\circ\text{C}$ $80\ ^\circ\text{C} \leq T_{\text{sink}} \leq 160\ ^\circ\text{C}$ $25\ \text{K} \leq \Delta T_{\text{lift}} \leq 95\ ^\circ\text{C}$	$a = 1.9118$ $b = 0.044189$ $c = -0.89094$ $d = 0.67895$

According to [8]

Closing the energy balance around the heat pump using Eq. (4) and with the help of Eqs. (2) and (3), all the necessary energy flows are obtained.

4 Evaluation

The purpose of this chapter is to evaluate the heat pump integration concept to be investigated in an industrial energy system. For this purpose, the evaluation principles and the parameters to be varied are presented in a first step. The results of the technical evaluation are then presented in the next section.

4.1 Basics

The evaluation of integrating a heat pump into the district heating circuit is carried out using technical key figures. Therefore, a heat pump with a thermal energy output of 1 MW is integrated into the system. The focus is on the change in energy consumption caused by integrating a heat pump into the district heating supply system. For this purpose, to investigate the integration's influence on the system's overall efficiency, the process parameter supply temperature of the heating system in the consumption circuit is varied. The values of the parameter variation are shown in Table 2.

The supply temperature of the system has been varied using standard temperatures of industrial heating systems. The reference system is the district heating supply with a consumption circuit supply temperature of 110 °C. The heat consumption of the reference system is 15,029 MWh per year.

4.2 Results

In order to evaluate the integration of the heat pump into the district heating supply, the supply temperature of the consumption circuit was varied according to the procedure in Table 2. The results of the simulations and the comparison towards the reference system are displayed in Table 3.

Table 2 Overview on varied process parameter

Process parameter	Simulation	Values (°C)
Supply temperature	1	100
	2	110
	3	120

Table 3 Simulation results for the variation of the supply temperature

Simulation	Thermal output DH	Thermal output HP	Electrical consumption HP	Annual COP	ω
Reference	15,039 MWh	0	0	0	0.99
1	14,130 MWh	3433 MWh	793 MWh	4.33	1.01
2	14,233 MWh	3477 MWh	691 MWh	5.03	1.01
3	14,257 MWh	3981 MWh	669 MWh	5.95	1.01

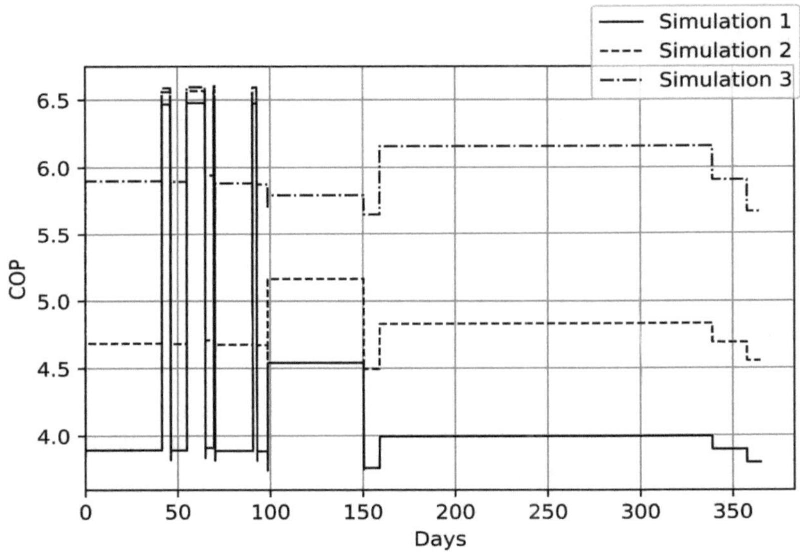

Fig. 3 Annual COP curves of the simulation configurations

As shown in the table above, a heat pump's integration increases the overall system efficiency ω. This is due to the operating principle of the heat pump. In addition, it can be seen that a higher flow temperature in the heating circuit leads to a higher COP on an annual basis. The higher flow temperature in the consumption circuit leads to a higher return temperature of the DH circuit. The return temperature of the DH circuit is equal to the heat pump source temperature, and thus a lower temperature lift is achieved. The average temperature lift within simulation 1 is equal to 40 K, within simulation 2 equal to 32.9 K and within simulation 3 equal to 25.9 K. Taking these results into account, it can be stated that the annual COP strongly depends on the supply temperature. The annual COP curves of the simulations are presented in Fig. 3. The figure also shows that the COP increases with the load. This is due to the fact that the system is designed for the maximum load.

5 Summary

This chapter investigates the integration of a heat pump into the district heating supply of an industrial energy system. To evaluate the heat pump integration, the supply temperature was varied as a process parameter. It was shown that the annual coefficient of performance of the heat pump increases with higher supply temperatures due to the return flow temperature of the DH circuit. In addition, integrating a heat pump increases the overall efficiency of the system and thus reduces the primary energy consumption. Furthermore, further analysis could include an economic comparison of the options. This would require the inclusion of the actual integration costs, which were not collected in this study.

Acknowledgement The authors gratefully acknowledge the financial support of the German Federal Ministry of Economic Affairs and Climate Action (BMWK), the project supervision of the Project Management Jülich (PtJ) for the project 'FlexGUIde', and the companies that provided their data for the purpose of this research.

References

1. European Commission. (2019). *The European Green Deal: Communication from the commission to the European parliament, the European Council, the Council, the European Economic and Social Committee and the Committee and the Commitee of the regions*. https://eur-lex.europa.eu/resource.html?uri=cellar:b828d165-1c22-11ea-8c1f-01aa75ed71a1.0002.02/DOC_1&format=PDF
2. Eurelectic. (2018). *Decarbonisation pathways: Full study results*.
3. Arpagaus, C., Bless, F., Uhlmann, M., et al. (2018). High temperature heat pumps: Market overview, state of the art, research status, refrigerants, and application potentials. *Energy, 152*, 985–1010. https://doi.org/10.1016/j.energy.2018.03.166
4. Kosmadakis, G. (2019). Estimating the potential of industrial (high-temperature) heat pumps for exploiting waste heat in EU industries. *Applied Thermal Engineering, 156*, 287–298. https://doi.org/10.1016/j.applthermaleng.2019.04.082
5. Arpagaus, C. (2019). *Hochtemperatur-Wärmepumpen: Marktübersicht, Stand der Technik und Anwendungspotenziale*. VDE Verlag GmbH.
6. Schneider, G. (Ed.). (1985). *Physik*. Trauner.
7. Wolf, S. (2017). *Integration von Wärmepumpen in industrielle Produktionssysteme: Potenziale und Instrumente zur Potenzialerschließung*. Universität Stuttgart.
8. Schlosser, F., Jesper, M., Vogelsang, J., et al. (2020). Large-scale heat pumps: Applications, performance, economic feasibility and industrial integration. *Renewable and Sustainable Energy Reviews, 133*, 110219. https://doi.org/10.1016/j.rser.2020.110219
9. Schlosser, F. (2021). *Integration von Wärmepumpen zur Dekarbonisierung der industriellen Wärmeversorgung*. https://kobra.uni-kassel.de/bitstream/123456789/12737/1/kup_9783737609425.pdf

Part V
Distributed Energy Resources Based Microgrid and Battery Energy Storage Technology

Assessing Economic Performance of an Energy Microgrid: A Conditional Value-at-Risk Optimization Approach

Seyedehsahar Seyedbarhagh, Hannu Laaksonen, and Mazaher Karimi

1 Introduction

Distributed energy resources (DERs) have gained significant prominence in power systems in recent years due to their ability to effectively mitigate greenhouse gas emissions while simultaneously offering power with a favorable penetration coefficient and exceptional reliability [1]. These sources play a pivotal role in reducing environmental pollution by minimizing the release of harmful gases, and they contribute to the overall sustainability of power generation. Moreover, their notable penetration coefficient ensures efficient utilization of power, while their remarkable reliability guarantees a consistent and uninterrupted electricity supply. The incorporation of distributed generation sources revolutionizes the transmission of energy across the power grid and enables consumers to exercise greater flexibility in their energy usage. This adaptation of decentralized production sources fundamentally alters the dynamics of energy distribution, facilitating a more efficient and adaptable system. Consequently, consumers are empowered to optimize their energy consumption patterns, making the most of the available resources while responding to their specific needs and preferences. This shift toward a more flexible and consumer-centric energy landscape enhances overall energy utilization and promotes a more sustainable and customer-driven approach. A novel hybrid fuel cell power generation system with high efficiency is proposed in [2]. The economic feasibility of MWe-scale systems using different fuels was assessed through economic modeling. The goal of this work was to develop the technology roadmap for this type of clean power system.

S. Seyedbarhagh (✉) · H. Laaksonen · M. Karimi
School of Technology and Innovations, Electrical Engineering, University of Vaasa, Vaasa, Finland
e-mail: seyedehsahar.seyedbarhagh@uwasa.fi

P. Pong (ed.), *Renewable Energy Resources and Conservation*, Green Energy and Technology, https://doi.org/10.1007/978-3-031-59005-4_25

The authors in Ref. [3] presented a pioneering contribution that introduces an optimal energy management strategy, focusing on identifying and utilizing the maximum efficiency range (MER) for a hybrid car equipped with both a fuel cell and a battery. The primary objective of this study is to address the challenges associated with energy management in such hybrid systems, aiming to optimize overall performance and efficiency. By strategically leveraging the MER, the proposed strategy aims to effectively manage power flow and utilization between the fuel cell and the battery, ensuring optimal utilization of both energy sources. This innovative approach contributes to the advancement of energy management strategies for hybrid cars, aligning with the broader objective of achieving sustainable transportation systems with enhanced energy efficiency. The authors in [4] focused on the design of an off-grid wind power generator with a hydrogen energy storage system. The authors propose a biologically inspired optimization algorithm that considers the total cost and load loss as objective functions. The algorithm, based on artificial bee colony optimization, outperforms the particle swarm optimization algorithm.

The reviewed studies highlight the significant advancements in the design, optimization, and integration of hybrid renewable energy systems for various applications. The findings demonstrate the potential of these systems in achieving cost-effective and sustainable energy generation, reducing carbon emissions, enhancing reliability, and improving the utilization of renewable resources. By considering the insights gained from these studies, future research can further advance the development and implementation of hybrid renewable energy systems, contributing to a cleaner and more resilient energy future.

In this chapter, we are studying an energy system that includes several DERs and planning to minimize the total cost of the energy system. The uncertainty posed by the electricity and gas prices is addressed using the CVaR method.

2 The Studied Model

The studied model including its components is illustrated in Fig. 1 [5]. The objective of the model is to minimize the overall cost of the energy system through the application of CVaR method. This cost encompasses the expenses associated with electricity procurement from the upstream network and gas procurement from the gas network, while also considering the revenue from selling power to the upstream network. The objective function is expressed in Eq. (1). By minimizing the total cost of the system, the model aims to optimize the utilization of available resources. This objective function serves as a guide for decision-making processes in determining the optimal configuration and operation of the hybrid system. To achieve this objective, various factors are considered, including the prevailing electricity and gas prices, the power output of the system, and the amount of gas consumed. Also, stochastic programming using the CVaR method is applied in this formulation [6].

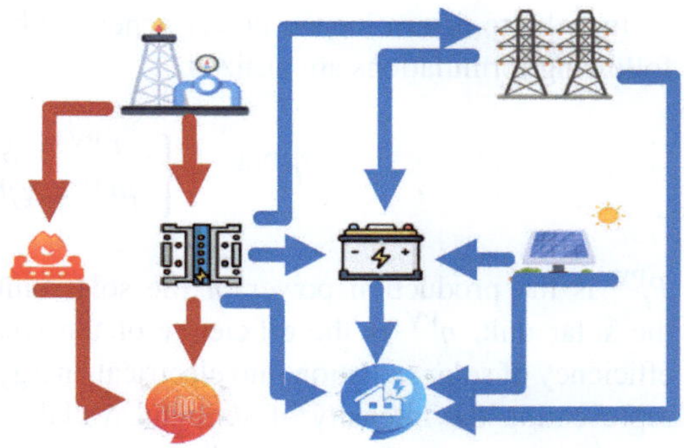

Fig. 1 The components of the studied energy system

$$\text{Min } (1-\beta)\left(\sum_{t=1}^{NT}\left[\sum_{sc=1}^{SC}\pi\,(\text{sc})\left(P_t^{\text{Net}}\lambda_{t,\text{sc}}^{\text{Net}}+G_t^{\text{Net}}\lambda_{t,\text{sc}}^{\text{Gas}}\right)\right]-P_t^{\text{Exp}}\lambda_t^{\text{Net,sell}}\right)$$
$$-\beta\left(\eta+\frac{1}{1-\alpha}\sum_{sc=1}^{SC}\pi\,(\text{sc})\,\xi_{\text{sc}}\right) \tag{1}$$

The probability of the scenarios is shown by $\pi(\text{sc})$. β is the risk level of the risk measure. It can be a value between 0 and 1. The higher values of β make the system more risk-averse. The variable P_t^{Net} in the context of this study represents the quantity of electrical power procured from the upstream network. By accurately quantifying the power purchased from the upstream network, the optimization process aims to minimize expenses and optimize the system's energy procurement. The parameter $\lambda_{t,\text{sc}}^{\text{Net}}$represents the price or cost associated with the power purchased from the upstream network at time t and scenario sc. The variable G_t^{Net} represents the amount of gas purchased from the gas network. Similarly, $\lambda_{t,\text{sc}}^{\text{Gas}}$represents the price of gas purchased from the gas network at time t and scenario sc. Furthermore, P_t^{Exp}introduced as an amount of power sold back to the upstream network. This variable accounts for any excess power generated by the system that can be supplied back to the upstream network. Finally, $\lambda_t^{\text{Net,sell}}$ is the selling electricity price to the network.

By balancing the power production and demand, the model aims to maximize the utilization of renewable energy sources while effectively meeting the energy needs of the system.

$$P_t^D+P_t^{\text{Exp}}+P_t^{\text{ESS},C}=P_t^{\text{Net}}+P_t^{\text{DER}}+P_t^{\text{ESS},D} \tag{2}$$

where variable P_t^D is the electric load, variable $P_t^{\text{ESS},C}$ quantifies the charging power of the storage. Additionally, P_t^{DER} encompasses the combined power generated by the fuel cell and solar cells, representing the local renewable energy production. $P_t^{\text{ESS},D}$ represents the power discharged from the storage system.

In order to determine the power generated by the solar system and fuel cell, the following formulations are utilized.

$$P_t^{\mathrm{DER}} = \begin{cases} P_t^{\mathrm{PV}} = S\eta^{pv} R_t \\ P_t^{\mathrm{FU}} = G_t^{\mathrm{FU}} \eta^{\mathrm{FU},E} \end{cases} \quad (3)$$

P_t^{PV} is the production power of the solar unit. S is the area required to install the solar unit. η^{PV} is the efficiency of the solar unit quantifying the conversion efficiency of solar radiation into electrical energy. R_t is the amount of solar radiation representing the intensity of sunlight available for power generation. P_t^{FU} is the production power of the fuel cell unit. G_t^{FU} is the amount of gas required for the operation of a fuel cell, typically hydrogen or a hydrogen-rich fuel, and $\eta^{\mathrm{FU},E}$ is the efficiency of the fuel cells.

In addition, it is essential to address the thermal aspect of the system, particularly the heat generated by local sources such as the fuel cell and backup boiler system, in order to meet the thermal load requirements.

$$H_t^D = H_t^{\mathrm{DER}} \quad (4)$$

The thermal load of the system is denoted as H_t^D. On the other hand, the heat produced by local sources such as fuel cell and backup boiler are represented by H_t^{DER}.

To accurately capture and model the heat generation from these sources, an equation is proposed, which describes the relationship between the produced heat and the operational parameters of the fuel cell and backup boiler system.

$$H_t^{\mathrm{DER}} = \begin{cases} H_t^B = G_t^B \eta^B \\ H_t^{\mathrm{FU}} = G_t^{\mathrm{FU}} \eta^{\mathrm{FU},H} \end{cases} \quad (5)$$

The production heat of the backup unit is denoted by H_t^B. Similarly, the production heat of the fuel cell is shown by H_t^{FU}.

The limits of energy production by local sources, which are crucial in understanding the system's capabilities and constraints, are detailed below. These limits represent the upper bounds of energy generation that can be achieved using the available local resources and technologies, providing valuable insights into the system's potential and operational limitations and ensure optimal energy production and utilization.

$$E_{\mathrm{Min}}^{\mathrm{DER}} \le E_t^{\mathrm{DER}} \le E_{\mathrm{Max}}^{\mathrm{DER}} \quad (6)$$

The power limit exchanged between the energy system and the upstream network are defined as follows:

$$P_{\mathrm{Min}}^{\mathrm{Net}} \le P_t^{\mathrm{Net}} \le P_{\mathrm{Max}}^{\mathrm{Net}} \quad (7)$$

$$P_{\text{Min}}^{\text{Exp}} \leq P_t^{\text{Exp}} \leq P_{\text{Max}}^{\text{Exp}} \tag{8}$$

The battery storage model is formulated as follow:

$$W_t^{\text{ESS}} = W_{t-1}^{\text{ESS}} + P_t^{\text{ch}}\eta^{\text{ch}} - P_t^{\text{dis}}\eta^{\text{dis}} \tag{9}$$

$$W^{\text{Min}} \leq W_t^{\text{ESS}} \leq W^{\text{Max}} \tag{10}$$

$$P_{\text{Min}}^{\text{ch}} I_t^{\text{ch}} \leq P_t^{\text{ch}} \leq P_{\text{Max}}^{\text{ch}} I_t^{\text{ch}} \tag{11}$$

$$P_{\text{Min}}^{\text{dis}} I_t^{\text{dis}} \leq P_t^{\text{dis}} \leq P_{\text{Max}}^{\text{dis}} I_t^{\text{dis}} \tag{12}$$

$$I_t^{\text{ch}} + I_t^{\text{dis}} \leq 1 \tag{13}$$

W_t^{ESS} is the energy stored in the battery. W_t^{ESS} represents the amount of energy that is currently stored in the battery. η^{ch} and η^{dis} are the charging and discharging efficiencies, respectively. W^{Min} and W^{Max} are the minimum and maximum energy stored in the battery, respectively. I_t^{ch} and I_t^{dis} are binary variables of the charging and discharging models.

$$\sum_{t=1}^{\text{NT}} \left[\left(P_t^{\text{Net}} \lambda_{t,\text{sc}}^{\text{Net}} + G_t^{\text{Net}} \lambda_{t,\text{sc}}^{\text{Gas}} \right) - P_t^{\text{Exp}} \lambda_t^{\text{Net,sell}} \right] - \eta \leq \xi_{\text{sc}} \tag{14}$$

$$\xi_{\text{sc}} \geq 0 \tag{15}$$

Constraints (14) and (15) represent the obligatory conditions for incorporating CVaR calculations within the stochastic programming model. As previously indicated, the application of the CVaR approach is proposed to evaluate the effect of the uncertain parameter. The electricity and gas purchased from the upstream network, i.e., $\lambda_{t,\text{sc}}^{\text{Net}}$ and $\lambda_{t,\text{sc}}^{\text{Gas}}$, are the uncertain parameters. α is a level of assurance, η is VaR, ξ_{sc} is the positive variable.

3 Result Discussions

In this section, a system has been employed for simulating the presented model, as depicted in Fig. 1. The daily operational cost of the energy system is \$44. This cost reduction is attributed to the optimal utilization of local systems such as fuel cells in energy production and their efficient use.

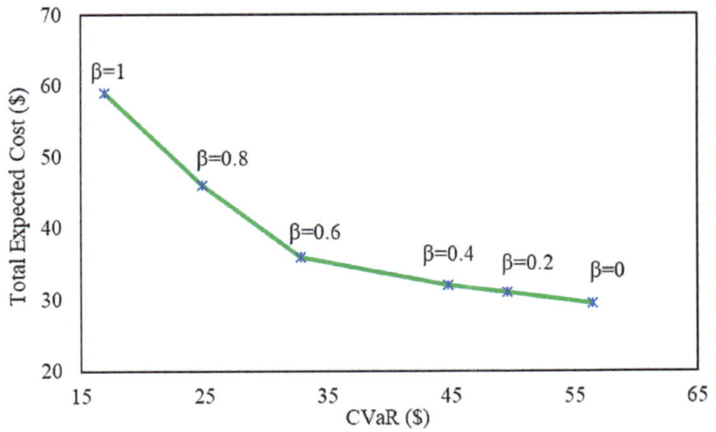

Fig. 2 The impact of the risk parameter on the total expected cost and CVaR

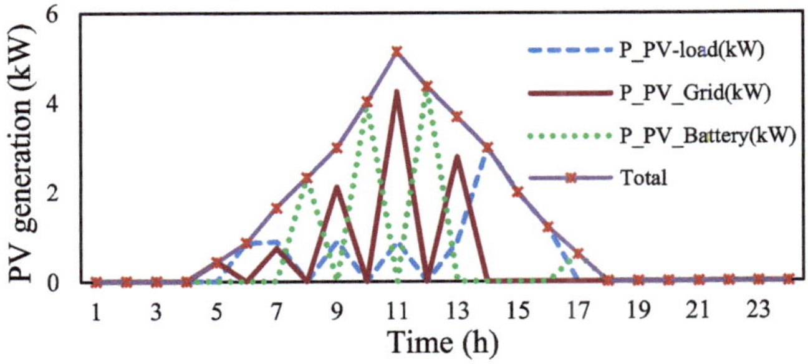

Fig. 3 Power produced by the photovoltaic system

Figure 2 presents a visual representation of the expected total cost changes across various values of CVaR. In this study, α is set to 0.95, and β ranges from 0 to 1, with increments of 0.2. When β equals 0, the operator exhibits a risk-averse behavior, leading to the highest possible CVaR. As β increases, the decision-maker becomes more accepting of increased risk when dealing with uncertain parameters. Additionally, raising the value of β leads to a decrease in CVaR, implying that as risk tolerance increases, operational costs are minimized.

The level of power generation by the solar unit is illustrated in Fig. 3. As observed, solar panel power generation commences at 5 AM and continues until 5 PM. The generated power is distributed to three units separately: battery, grid, and load. The highest solar panel power generation occurs in the afternoon, and this excess power is either sold back to the grid upstream or stored in the battery unit.

4 Conclusions

In this chapter, the problem of optimal operation of an energy system is examined through application of CVaR as the risk management method. The goal of the proposed model is to optimize the utilization of local resources for energy supply and achieve economic benefits, resulting in minimizing the operational costs of the system. According to the results, it is shown that as the risk level increases, the operator becomes more accepting of increased risk when dealing with uncertain parameters. In future work, the possibility of considering electrical and gas network models can be explored.

References

1. Ahmadi, B., Ceylan, O., Ozdemir, A., & Fotuhi-Firuzabad, M. (2022). A multi-objective framework for distributed energy resources planning and storage management. *Applied Energy, 314*, 118887.
2. Yan, H., et al. (2020). Techno-economic evaluation and technology roadmap of the MWe-scale SOFC-PEMFC hybrid fuel cell system for clean power generation. *Journal of Cleaner Production, 255*, 120225.
3. Wang, T., et al. (2020). An optimized energy management strategy for fuel cell hybrid power system based on maximum efficiency range identification. *Journal of Power Sources, 445*, 227333.
4. Wang, R., & Zhang, R. (2023). Techno-economic analysis and optimization of hybrid energy systems based on hydrogen storage for sustainable energy utilization by a biological-inspired optimization algorithm. *Journal of Energy Storage, 66*, 107469.
5. Majidi, M., Nojavan, S., & Zare, K. (2017). Optimal stochastic short-term thermal and electrical operation of fuel cell/photovoltaic/battery/grid hybrid energy system in the presence of demand response program. *Energy Conversion and Management, 144*, 132.
6. Guo, Q., Nojavan, S., Lei, S., & Liang, X. (2021). Economic-environmental analysis of renewable-based microgrid under a CVaR-based two-stage stochastic model with efficient integration of plug-in electric vehicle and demand response. *Sustainable Cities and Society, 75*, 103276.

Study of the Behavior of an Electric Power Generation System with AGM Battery Storage Using Sankey Diagrams

Andrés Felipe Parada Valle and Fabio Emiro Sierra Vargas

1 Introduction

Around the world, 17% of the population have no access to electricity. Nearly 80% live in rural areas. A low electrification constrains agricultural production due a lack of energy needed for post-harvesting processing that requires cooling and heating devices [1]. Electrical generators powered by fuels are available. Nevertheless, Gasoline or Diesel are commonly used and cause negative impacts on the environment [2]. The use of natural gas and biogas generators is an attractive alternative option to obtain electricity from fuels with lesser emissions [3]. Energy storage allows excess energy to be captured and improves system efficiency by increasing energy utilization [4].

2 Methodology and Equipment

A first experiment was carried out on a 110 V AC generator under varying loads to determine its efficiency curve. Then, a second experiment was performed on a setup containing a 110 V Generator, an AC-DC Transformer, and two different 12 V Batteries. The aim is to illustrate, using Sankey diagrams, the magnitude of losses in each utilized equipment until the energy is received and stored by the battery. A description of each equipment and the equations used are shown. The laboratory equipment was properly instrumented for measuring fuel consumption FC, current, and voltage. Three battery charging cycles were conducted and data is being averaged. The outcomes are presented using Sankey diagrams at 50%

A. F. Parada Valle (✉) · F. E. Sierra Vargas
Universidad Nacional de Colombia, Bogota D.C, Colombia
e-mail: aparadav@unal.edu.co; fesierrav@unal.edu.co

P. Pong (ed.), *Renewable Energy Resources and Conservation*, Green Energy and Technology, https://doi.org/10.1007/978-3-031-59005-4_26

and 100% charge levels to ascertain the influence of the charge percentage on the system's efficiency.

2.1 Fuel-Based Electric Generator Balance

This allows conversion of fuel-based power into electric power. The energy balance is given by the first law of thermodynamics that describes the change in the internal energy of the system as proposed in [5]:

$$\sum Q = \sum H_{\text{in}} - H_{\text{out}} = Q_{\text{in}} = W_{\text{u}} + H_{\text{looses}} \tag{1}$$

The heat input Expressed in terms of the Lower Heating Value (LHV) and the mass flow of fuel, which can be also evaluated assuming the fuel as an ideal gas, as shown.

$$Q_{\text{in}} = Q_{\text{comb}} = \dot{m}_{\text{fuel}} \cdot \text{LHV} = \frac{m}{\Delta t} \cdot \text{LHV} = \frac{\text{PV}}{RT\Delta t} \cdot \text{LHV} = \frac{\rho V}{\Delta t} \cdot \text{LHV} \ [W] \tag{2}$$

Power output, given as useful work, represents the amount of the energy that is converted into electricity, and can be calculated by the following equation

$$W_{\text{u}} = V_{\text{AC}} \cdot I_{\text{AC}} \ [W] \tag{3}$$

Efficiency: The efficiency of the generator is the ratio between the energy released by the fuel and the energy actually used as electricity. In terms of percentage is given by

$$\eta_{\text{g}} = \frac{W_u}{Q_{\text{in}}} \cdot 100 \ [\%] \tag{4}$$

Shenzen Puxin PX3600 EGB 3000 W Natural Gas/Biogas Generator: Comprises a Honda GX390 ICE [6] fueled by natural gas and a single-phase brush motor [7], providing an AC output of 110 V. Properties of Colombian gas were taken from previous research [8].

$$\text{LHV} = 47.645 \ \text{kJ/g} \tag{5}$$

$$P = 522.0363 \ \text{kg/m}^3 \tag{6}$$

2.2 Load Bank

Load Banks are used to generate testing loads for power supplies without connecting it to its normal operating load. The bank used has six different loads allowing up to nearly 2 kW. The magnitude of the load is measured as follows:

$$W_{\mathrm{RB}} = V_{\mathrm{RB}} \cdot I_{\mathrm{RB}} \; [W] \tag{7}$$

2.3 110 V AC to 12 V DC Transformer

An AC-DC transformer has a 110 V AC input, to obtain a direct current at 12 V. Due to the nature of the components, resistance, transformation, and rectification, losses are expected affecting the efficiency of the process. The efficiency can be calculated as

$$\eta_{\mathrm{TR}} = \frac{V_{\mathrm{DC}} \cdot I_{\mathrm{DC}}}{V_{\mathrm{AC}} \cdot I_{\mathrm{AC}}}. \tag{8}$$

Schumacher SC1309 Battery Charger: This equipment allows a transformation and rectification from 110 V AC to 12 V DC at a current rate of 2 A or 40 A. The experiment was conducted using the 40 A mode [9].

2.4 Energy Storage Device

The optimal solution for off-grid hybrid renewable system is usually a *PV + Fossil fuel generator + battery energy storage*. Lead-acid battery is one of the most used.

MTEK MT123000 Valve Regulated Lead Acid Battery is a high-performance power solution with a rating of 3600 W operating at 12 V. Its robust design ensures reliable performance [10]. Nevertheless, deep cycling discharge affects performance and lifespan. An equivalence between percentage of charge and voltages were obtained through battery data and experimentation.

3 Results

Graphing the voltage, current, and efficiency of the genset obtained in Table 2.

Representing this results by means of Sankey Diagrams, the following figure is obtained

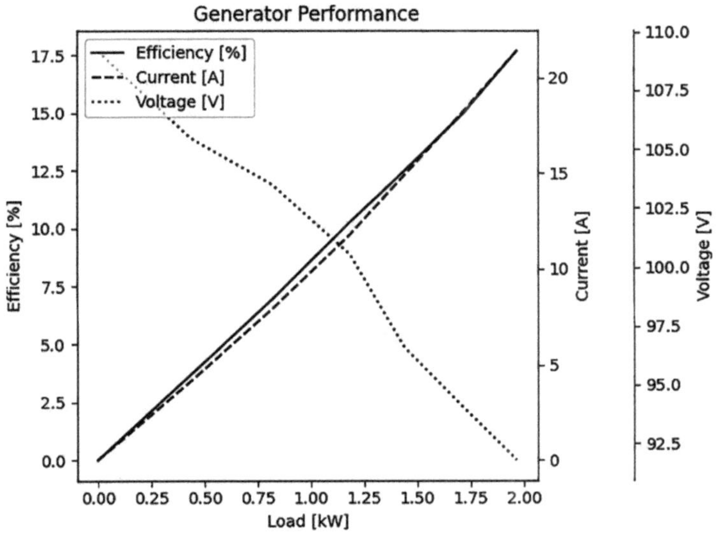

Fig. 1 Generator performance curves

Table 1 Percentage of charge and voltages equivalences

Percentage of charge [%]	Resting voltage [V]	Charging voltage [V]
50%	12.10	12.70
60%	12.24	12.86
70%	12.37	13.02
80%	12.50	13.18
90%	12.62	13.34
100%	12.73	13.5

4 Discussion

Fuel consumption remains nearly constant regardless of load, indicating a lack of fuel regulation, possibly due to the generator's size. Equation 2 relates pressure and gas mass via the ideal gas law. Adding pressure regulation to the gas line could reduce consumption, enhancing economy. Useful work rises, increasing efficiency at higher loads (Fig. 1). The Sankey diagram depicts generator energy loss magnitude. Table 1 shows a load of 0.2–0.4 kW on the generator during the charging process, aligning with low efficiency range (Table 2). Charger maintains ~58% efficiency but current varies with charge percentage, staying below 40A and decreasing. Figure 2 highlights the system losses, revealing the proportion of losses at each stage of the process. A greater significance of the loss in the motor-generator is identified, as well as a loss of about half of the energy received by the charger, leading to very low overall efficiency during the charging process (Table 3).

Table 2 Generator efficiency results

Load	FC [m^3/min]	$\dot{m}_{fuel}$[g/s]	Q_{in}[kW]	V_{AC}[V]	I_{AC}[A]	W_U[kW]	η_G[%]
1	0.028	0.247	11.77	105.5	4.1	0.432	3.674
2	0.028	0.247	11.77	103.5	7.9	0.817	6.945
3	0.028	0.247	11.77	100.5	11.8	1.186	10.290
4	0.028	0.247	11.77	96.6	14.9	1.439	12.489
5	0.028	0.247	11.77	94.0	18.3	1.720	15.034
6	0.028	0.247	11.77	91.8	21.4	1.964	17.682

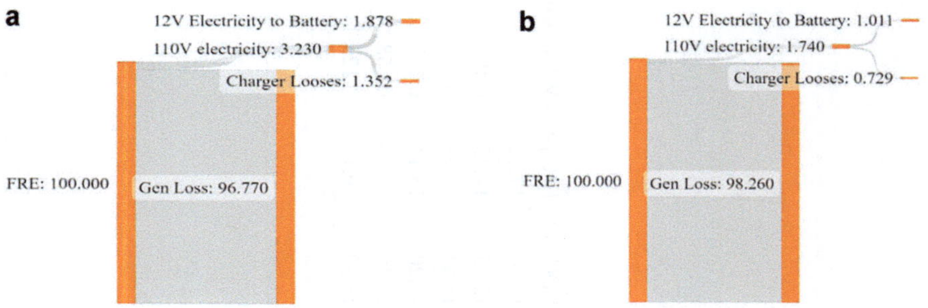

Fig. 2 (**a**) Sankey Diagram 50% of charge (**b**) Sankey Diagram 100% of charge

5 Conclusions

These experiments demonstrated that as the system nears nominal power, it becomes more efficient, yet there's no fuel regulation. Since the gas acts as an ideal gas, this energy loss could be related to the absence of a pressure-regulating valve. When charging batteries, the requested load is much lower than the engine's nominal load, so an oversized engine with pressure regulation would be unnecessary unless load capacity is accurately determined for battery charging while maintaining peak efficiency. Future studies are recommended to use a pressure-regulating valve or a regulated generator to analyze system efficiency by reducing fuel mass. The energy delivered to the load is even lower due to charger losses, which remained at about 58% efficiency and approximately constant. The charger's internal reason for losing energy isn't clear, but it's evident that it delivers less current as batteries charge, consistent with an AGM battery's charging curve. This affects the power demanded from the generator, making the engine less efficient as the battery nears full charge. These recommendations are expected to lead to more economically feasible models by reducing gas consumption during the charging process.

Table 3 Charging process results

Charge [%]	FC [m^3/min]	$\dot{m}_{fuel}$[g/s]	Q_{in}[kW]	V_{AC}[V]	I_{AC}[A]	W_U[kW]	G[%]	VDC [V]	IDC [A]	WBat [kW]	Charg [%]	Total [%]
50%	0.028	0.247	11.60	103.10	3.64	0.374	3.230	12.70	17.17	0.217	58.14	1.878
60%	0.029	0.252	12.02	101.95	3.20	0.326	2.714	12.85	14.85	0.190	58.44	1.587
70%	0.029	0.252	12.02	101.50	3.08	0.312	2.590	13.02	13.98	0.181	58.31	1.510
80%	0.029	0.252	12.02	101.60	300	0.304	2.511	13.20	13.11	0.172	58.84	1.427
90%	0.029	0.252	12.02	101.85	2.40	0.244	2.005	13.35	11.86	0.158	58.45	1.298
100%	0.029	0.252	12.02	101.50	2.10	0.213	1.749	13.50	9.13	0.123	58.36	1.011

References

1. Worldbank. (2015). *Electrifying rural areas*.
2. Llanes, E. (2018). *Energy and exergy evaluation in a 1.6L Otto cycle internal combustion engine*. SEK International University.
3. Dufo, R. (2022). Off-Grid full renewable hybrid systems: Control strategies, optimization, and modeling. In *Hybrid technologies for power generation*. Elsevier Science & Technology.
4. Díaz, I. (2018). *Análisis energético, exergético y medioambiental de la hibridación con energía solar de una planta de cogeneración*. Universidad Politécnica de Madrid.
5. Cengel, Y. (2019). *Thermodynamics: An engineering approach* (9th ed.). McGrawHill.
6. Honda, Honda GX390 Manual.
7. Shenzhen USG Technology Co., Ltd.
8. Franco, D. *Caracterización de un motogenerador de 3.5kW operado con biogás y almacenamiento en baterías AGM* (MsC Thesis). Universidad Nacional de Colombia.
9. Shumacher Electric, SC1309 Automatic Battery Charger, owner's manual.
10. MTEK, Technical datasheet for MT123000 12V 300AH Battery.

Electro-acoustic Charging Prolongs the Cycle Life of Lead-Acid Battery Cells

Drandreb Earl O. Juanico

1 Introduction

The rise of electric vehicles (EVs) is poised to leave behind 1.7 million metric tons of starter lighting and ignition (SLI) batteries, essentially flooded lead-acid batteries (LABs) [1]. Historically, LAB production emphasizes sustainability, with a notable 90% of SLI LABs being recycled due to strict environmental regulations [2]. This recycling rate positions LABs uniquely, embodying circular economy principles [3].

Nevertheless, the automotive industry's focus is shifting from LABs toward alternatives like lithium-ion batteries [1]. As EVs become mainstream, a vast network of charging stations, especially off-grid ones, will emerge. These stations require reliable stationary energy storage, which LABs can potentially provide.

Repositioning LABs for this purpose aligns with the world's move toward EVs, emphasizing the need for efficient energy storage amid renewable energy's intermittent nature. LABs could serve not only EV charging stations but also off-grid DC-powered homes [4, 5].

However, maximizing LABs' potential means addressing their limited cycle life, especially under intermittent charging, and other challenges like sulfation [6]. Our study introduces "electro-acoustic charging" for LABs, a method combining electricity and sound, tailored for renewable energy storage. Our findings suggest this approach could extend LABs' cycle life.

The original version of the chapter has been revised. A correction to this chapter can be found at https://doi.org/10.1007/978-3-031-59005-4_28

D. E. O. Juanico (✉)
Advanced Batteries Center, Quezon City, NCR, Philippines

EEEI, University of the Philippines, Diliman, Quezon City, Philippines
e-mail: dojuanico@up.edu.ph

P. Pong (ed.), *Renewable Energy Resources and Conservation*, Green Energy and Technology, https://doi.org/10.1007/978-3-031-59005-4_27

2 Materials and Methods

Cells were sourced from a motorcycle SLI flooded LAB, specifically the Motolite® 12 N12 model. These cells have a capacity of 2 Ah each and operate at 12 V. They were submerged in a sulfuric acid solution with a specific gravity of 1.25, indicating 30–34% acid by volume. An absorbent glass mat separator, crucial for preventing short circuits and facilitating electrochemical reactions, was placed between the cell electrodes. The cell compartment, made of polypropylene, was retained from the original battery.

For instrumentation, a piezoelectric transducer with a 1.7 MHz ultrasonic frequency was positioned beneath the cell compartment, anchored with a 3D printed thermoplastic attachment (see Fig. 1). An infrared (IR) thermal imager was set up facing the cell, supported by an LCD for real-time thermal monitoring.

Cells underwent charging and discharging cycles at a C-rate of C/4, managed by a BioLogic VSP-3e potentiostat. Charging terminated at 2.40 V or 4 h, and discharging at 1.75 V or the 4-h limit. An IR camera documented temperature variations, while two video cameras captured thermal and visual data.

The experiment involved a piezoelectric transducer, activated by a 60 V sinusoidal generator and an inductor-MOSFET circuit board, producing mechanical vibrations during the charging phase, termed "electro-acoustic" charging.

The study involved control samples and electro-acoustic samples. The state of health (SOH) of cells was monitored until it fell below 80%, indicating the

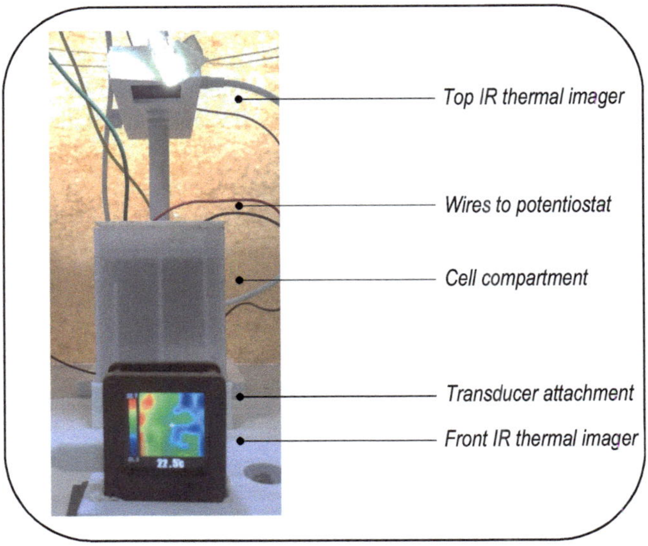

Fig. 1 Experimental setup connected to a potentiostat (not shown) with lead-acid cell and electrolyte inside the compartment and transducer attachment at the bottom of the compartment (partially shown)

cell's end of life. Post-experiment, electrodes and electrolytes were examined for changes. Additionally, simulations visualized internal pressure waves, enhancing the understanding of electro-acoustic charging and its effect on cell health.

3 Results and Discussion

The results presented herein represent a significant advancement in our understanding of lead-acid cell cycle life management. According to established research [7, 8], a lead-acid cell devoid of sound input, henceforth referred to as the control cell, would typically be limited to an average of four deep cycles under accelerated aging conditions (100% depth of discharge at a high C-rate). Contrary to this established knowledge, our data, particularly as delineated in Fig. 2, demonstrate that cells subjected to sound input, termed "experimental cells," exhibited a cycle life extending to over twice that of the control counterpart. Such a pronounced enhancement underscores the integral role of sound in augmenting the longevity of a cell.

Upon further investigation into the underlying mechanisms, it was ascertained that sound augments electrochemical reactions within the cell, consequently forestalling the onset of sulfation-induced failure, a prevalent mode of cell degradation. It is also noteworthy, as presented in Fig. 3, that the introduction of sound culminates in an elevation of temperature. However, this increase remains within permissible limits, ensuring that the electrolyte remains below boiling point and negating risks associated with active material displacement from the electrode grid or accelerated grid corrosion. This temperature escalation is posited to be intrinsically linked to the propagation of pressure waves.

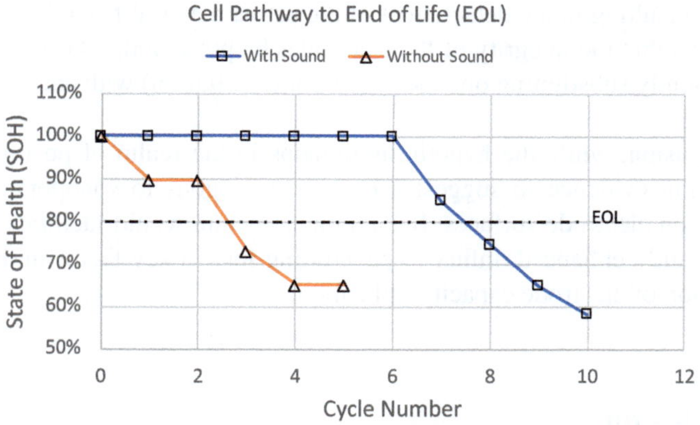

Fig. 2 Comparison of typical cycle life with and without sound (control) measured as the cycle number reached before SOH falls below EOL (dashed horizontal line)

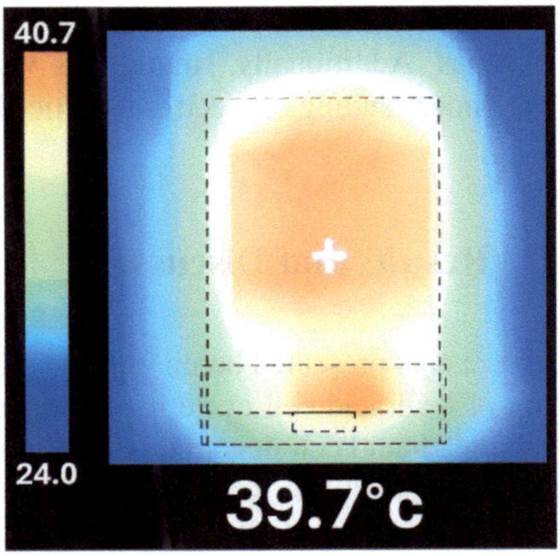

Fig. 3 Temperature distribution showing the outline (dashed) of the cell as viewed by the thermal imager from the front

A nuanced observation of these pressure waves revealed patterns of compression and rarefaction, which appear to modulate the distribution of pore pressure on the electrode surfaces. In reference to the patent application WO2022/191721 [9], it is postulated that this altered distribution enhances the permeation of electrolyte ions into the electrode matrix. In such a setting, these ions are afforded an elevated likelihood of interfacing with electrons from the current input. This mechanism could elucidate the observed enhancement in cell capacity utilization, which directly translates to an extended cycle life.

Our simulation data, elucidated in Fig. 4, further substantiates this hypothesis. The simulations indicate that, despite the critical role of pressure, it does not approach magnitudes that might precipitate phenomena such as inertial acoustic cavitation or the abrupt collapse of bubbles. Analogous pressures have been documented to induce pores on biological cells—a phenomenon termed "sonoporation" [10]. In the context of our study, excessive pressure could be deleterious, potentially resulting in the detachment of the active material from the electrodes. It is imperative that the integrity of the electrodes be preserved, otherwise, we would be inadvertently substituting one mode of failure (sulfation) with another (electrode disruption).

In conclusion, while the hypothesis remains in the realm of postulation, there is compelling evidence to suggest a process analogous to sonoporation may be transpiring on electrode surfaces. Such a phenomenon would facilitate a transient, but significantly enhanced, influx of penetrating ions, a key factor in the observed augmentation of electrode capacity utilization.

4 Conclusion

In this groundbreaking study, we have unveiled the potential of electro-acoustic charging in extending the cycle life of flooded lead-acid batteries (LABs). This

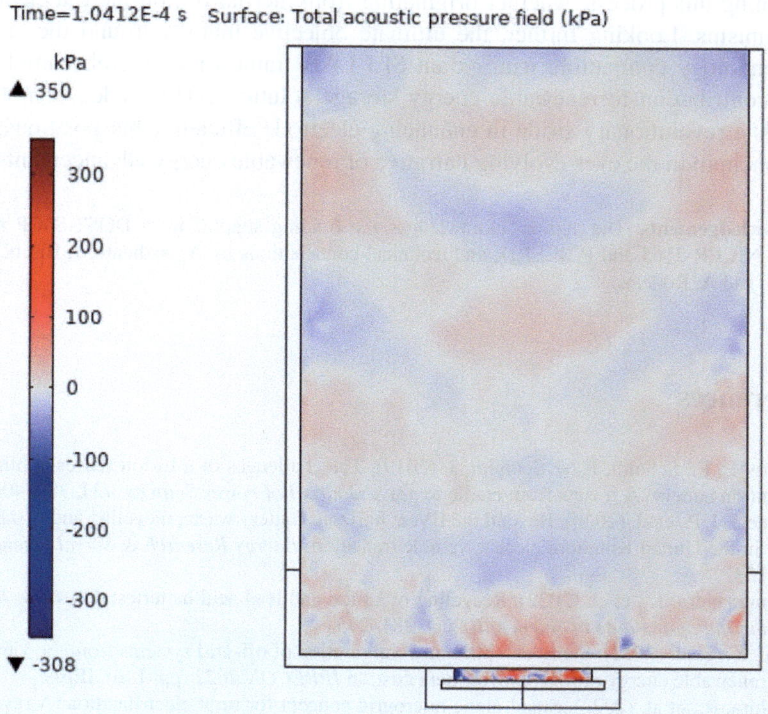

Fig. 4 Acoustic pressure waves visualized from a multi-physics simulation of the experimental cell subjected to 1.7 MHz sound input from the bottom

enhancement not only has the potential to significantly reduce the levelized cost of storage but also elevates the economic attractiveness of flooded LABs for stationary energy storage in off-grid settings. As the world seeks sustainable energy solutions, off-grid renewable energy systems equipped with LABs—currently the pinnacle of sustainable battery technology—could take center stage in the global energy transition. While our findings are promising, further investigations are crucial, including direct pressure measurements and spectral analysis of acoustic cavitation noise. Through continuous innovation and exploration, we believe in shaping a future where renewable energy solutions are not just an alternative, but the norm.

In light of recent exploratory investigation, future strides in electro-acoustic charging research brim with potential, foregrounding a two-pronged pathway for ongoing inquiry. Firstly, substantiating the observed temperature rise with a corresponding increase in gauge pressure would cement the pivotal role of acoustic pressure waves in optimizing ionic pore penetration through non-inertial cavitation, a cornerstone process that stands to markedly elevate the charge capacity of electrodes. Furthermore, the utilization of hydrophones for direct cavitation spectrum observation stands as a promising avenue for discerning the dynamics governing this process, whether originating from inertial or non-inertial cavitation mechanisms. Looking further, the ultimate objective merges around the crafting of technology compatible with extant SLI LAB frameworks, thereby amplifying their contribution to renewable energy storage solutions. This endeavor not only signals a revolutionary stride in enhancing electrode efficiency but positions itself as a linchpin in the ever-evolving narrative of renewable energy advancements.

Acknowledgements The author acknowledges the funding support from DOST-S4CP #2021-03-A1-NICER-3165 and PCIEERD; and technical contributions by A. Ardiente, J. Biscocho, N. Pacleb, and A. Rodelas.

References

1. Ferg, E. E., Schuldt, F., & Schmidt, J. (2019). The challenges of a Li-ion starter lighting and ignition battery: A review from cradle to grave. *Journal of Power Sources, 423*, 380–403.
2. Skeete, J. P., et al. (2020). Beyond the EVent horizon: Battery waste, recycling and sustainability in the United Kingdom electric vehicle transition. *Energy Research & Social Science, 69*, 101581.
3. Yanamandra, K., et al. (2022). Recycling of Li-ion and lead acid batteries: A review. *Journal of the Indian Institute of Science, 102*(1), 281–295.
4. Liu, K., et al. (2022). Study on optimum configuration of off-grid systems from the viewpoint of renewable energy ratio and investment cost. In *IPRECON 2022* (pp. 1–6). IEEE.
5. Zahira, R., et al. (2022). Stand-alone microgrid concept for rural electrification: A review. In *Residential microgrids and rural electrifications* (pp. 109–130). Academic Press.
6. Varshney, K., et al. (2020). Current trends and future perspectives in the recycling of spent lead acid batteries in India. *Materials Today: Proceedings, 26*, 592–602.
7. Blank, T., et al. (2012). Deep discharge behavior of lead-acid batteries and modeling of stationary battery energy storage systems. In *INTELEC 2012* (pp. 1–4). IEEE.
8. Thaller, L. H. (1983). Expected cycle life vs. depth of discharge relationships of well-behaved single cells and cell strings. *Journal of the Electrochemical Society, 130*(5), 986.
9. Juanico, D. E. (2022). Methods and system of acoustically assisted battery operation. PCT Application WO 98(21120), A1, A4.
10. Helfield, B., et al. (2016). Biophysical insight into mechanisms of sonoporation. *Proceedings of the National Academy of Sciences, 113*(36), 9983–9988.

Correction to: Electro-acoustic Charging Prolongs the Cycle Life of Lead-Acid Battery Cells

Drandreb Earl O. Juanico

Correction to:
Chapter 27 in: P. Pong (ed.), *Renewable Energy Resources and Conservation*, Green Energy and Technology,
https://doi.org/10.1007/978-3-031-59005-4_27

The updated affiliation for author Drandreb Earl O. Juanico in Chapter 27 was available in the Print versions, but not in the Online version. The affiliation was updated in the Online version.

The updated version of this chapter can be found at
https://doi.org/10.1007/978-3-031-59005-4_27

P. Pong (ed.), *Renewable Energy Resources and Conservation*, Green Energy and Technology, https://doi.org/10.1007/978-3-031-59005-4_28

Index

P. Pong (ed.), *Renewable Energy Resources and Conservation*, Green Energy and Technology, https://doi.org/10.1007/978-3-031-59005-4

MIX
Papier aus verantwortungsvollen Quellen
Paper from responsible sources
FSC® C105338

If you have any concerns about our products,
you can contact us on
ProductSafety@springernature.com

In case Publisher is established outside the EU,
the EU authorized representative is:
Springer Nature Customer Service Center GmbH
Europaplatz 3, 69115 Heidelberg, Germany

Printed by Libri Plureos GmbH
in Hamburg, Germany